NATO ASI Series

Advanced Science Institutes Series

A series presenting the results of activities sponsored by the NATO Science Committee, which aims at the dissemination of advanced scientific and technological knowledge, with a view to strengthening links between scientific communities.

The Series is published by an international board of publishers in conjunction with the NATO Scientific Affairs Division

A Life Sciences	Plenum Publishing Corporation
B Physics	London and New York
C Mathematical and Physical Sciences	Kluwer Academic Publishers Dordrecht, Boston and London
D Behavioural and Social Sciences	
E Applied Sciences	
F Computer and Systems Sciences	Springer-Verlag Berlin Heidelberg New York London Paris Tokyo Hong Kong Barcelona
G Ecological Sciences	
H Cell Biology	

The ASI Series Books Published as a Result of Activities of the Special Programme on SENSORY SYSTEMS FOR ROBOTIC CONTROL

This book contains the proceedings of a NATO Advanced Research Workshop held within the activities of the NATO Special Programme on Sensory Systems for Robotic Control, running from 1983 to 1988 under the auspices of the NATO Science Committee.

The books published so far as a result of the activities of the Special Programme are:

Vol. F25: Pyramidal Systems for Computer Vision. Edited by V. Cantoni and S. Levialdi. 1986.

Vol. F29: Languages for Sensor-Based Control in Robotics. Edited by U. Rembold and K. Hörmann. 1987.

Vol. F33: Machine Intelligence and Knowledge Engineering for Robotic Applications. Edited by A. K. C. Wong and A. Pugh. 1987.

Vol. F42: Real-Time Object Measurement and Classification. Edited by A. K. Jain. 1988.

Vol. F43: Sensors and Sensory Systems for Advanced Robots. Edited by P. Dario. 1988.

Vol. F44: Signal Processing and Pattern Recognition in Nondestructive Evaluation of Materials. Edited by C. H. Chen. 1988.

Vol. F45: Syntactic and Structural Pattern Recognition. Edited by G. Ferraté, T. Pavlidis, A. Sanfeliu and H. Bunke. 1988.

Vol. F50: CAD Based Programming for Sensory Robots. Edited by B. Ravani. 1988.

Vol. F52: Sensor Devices and Systems for Robotics. Edited by A. Casals. 1989.

Vol. F57: Kinematic and Dynamic Issues in Sensor Based Control. Edited by G. E. Taylor. 1990.

Vol. F63: Traditional and Non-Traditional Robotic Sensors. Edited by T. C. Henderson. 1990.

Vol. F64: Sensory Robotics for the Handling of Limp Materials. Edited by P. M. Taylor. 1990.

Series F: Computer and Systems Sciences, Vol. 64

Sensory Robotics for the Handling of Limp Materials

Edited by

Paul M. Taylor

Department of Electronic Engineering
University of Hull, Hull HU6 7RX, United Kingdom

Springer-Verlag Berlin Heidelberg New York
London Paris Tokyo Hong Kong Barcelona
Published in cooperation with NATO Scientific Affairs Division

Proceedings of the NATO Advanced Research Workshop on Sensory Robotics for the Handling of Limp Materials held in Il Ciocco, Italy, October 16–22, 1988.

ISBN-13: 978-3-642-75535-4 e-ISBN-13: 978-3-642-75533-0
DOI: 10.1007/978-3-642-75533-0

Softcover reprint of the hardcover 1st edition 1990

2145/3140-543210 – Printed on acid-free-paper

PREFACE

Limp materials are used in many economically important industries such as garment manufacture, shoe manufacture, aerospace (composites) and automobiles (seats and trim). The use of sensors is essential for reliable robotic handling of these materials, which are often based on naturally occurring substances such as cotton and leather. The materials are limp and have non-homogeneous mechanical properties which are often impossible to predict accurately. The applications are very demanding for vision and tactile sensing and signal processing, adaptive control systems, planning and systems integration.

This book comprises the collection of papers presented at the NATO Advanced Research Workshop on 'Sensory Robotics for the Handling of Limp Materials', held in October 1988 at Il Ciocco, Tuscany, Italy. The aim of the workshop was to examine the state of the art and determine what research is needed to provide the theoretical and technological tools for the successful application of sensory robotics to the handling of limp materials. The meeting also acted as the first-ever forum for the interchange of knowledge between applications-driven researchers and those researching into the provision of fundamental tools. The participants were drawn from academia (20), industry (5), and other non-university research organisations (5).

The organisation of the book broadly follows the workshop programme. The first half concentrated on the applications, looking at the particular problems of handling shoes, composites and garments, with an update on the state of applications-driven research wherever possible. Two round-table sessions were organised to try to identify trends and future needs. Various overview papers were then given on the state of 2D and 3D vision, tactile sensing, planning and control. Again, round-table discussions were held to discuss future needs and possibilities.

It will be seen that most current limp-material handling research is concerned only with 2D assemblies, and 2D vision techniques are used extensively in laboratories. However, there is an immediate industrial requirement for the development of low cost, highly reliable systems. Inspection systems for detecting material faults and errors in assemblies were seen as a great need - but again, low cost and high performance are necessary. New tactile sensors are wanted, for example to aid guidance of fabric during sewing and edge alignment operations. Many objects are three-dimensional in form, and this jump from 2D to 3D will provide a major future challenge.

Shoe and garment manufacture are very dependent on fashion and so reprogramming and reconfiguration of sensory assembly cells must make place in minutes, not weeks. It is important to address issues such as when to plan and when to sense. The properties of the materials need to be better understood, in a meaningful and quantified way. There are also possible applications for adaptive controllers which are designed to be tolerant to uncertainties in material properties.

The use of sensory robotics for the handling of limp materials is a new research area with a growing number of active research teams. As will be seen from these pages, there are many real and challenging problems for the sensory robotics community to address.

ACKNOWLEDGEMENTS

The editor would like to acknowledge the assistance given by Frank Paul, Holger Nollek and Clive Walter, particularly during the formative stages of the programme. Thanks are also extended to all other lecturers and participants for their presentations and for the free, wide-ranging discussions which formed such a feature of the workshop.

Hull, UK, January 1990

Paul Taylor

TABLE OF CONTENTS

III. Sensing

IV. Planning and Control

CHAPTER I

SHOES AND COMPOSITES

SENSORY ROBOTICS RELATED TO SHOE MANUFACTURE

P J Perkins
SATRA Footwear Technology Centre
SATRA House
Rockingham Road
Kettering, Northants, NN16 9JH
England

This paper reviews the materials and handling systems used in footwear manufacture and identifies the needs of the industry in order to move towards more automated techniques. A robot work cell for attaching shoe soles is described.

Need for flexible automation

Footwear manufacture has moved from a craft based industry to a more mechanised production process, but it remains highly labour intensive. Many of the critical assembly and processing operations still rely on manual movement of components and shoes from storage, or transportation medium, to machine and back as well as manipulation during the process.

For the developed countries it is becoming increasingly difficult to compete with imports from low labour cost areas on price terms, but the introduction of new technology such as CAD/CAM, automation, robotics etc is seen as one of the key ways in which the European and North American shoe industries can become more competitive. It will help by reducing labour costs and by increasing speed of response to retail requirements.

An increasing number of processes are being subjected to the application of new technology; handling systems using robots are being used in some companies. However, these are primarily where long production runs on the same shoe style are possible and mainly involve direct moulding of the sole to the upper.

NATO ASI Series, Vol. F 64
Sensory Robotics for the Handling of Limp Materials
Edited by P. M. Taylor

One of the key problems for the average shoe manufacturer is the number of variables in shoe production. There are a large number of shoe styles, different upper and soling material combinations, a large number of sizes and width fittings, plus the fact that each pair of shoes requires a left and a right. This problem is most prominent in shoe factories which produce women's fashion footwear. An average sized company will produce about 6,000 pairs of shoes per week. Typically, this would be made up of mixed pairages of 30 different styles of shoe in 11 different sizes and could even have 2 or 3 different width fittings. Moreover, the constraints of production line balancing mean that it is impossible to process all shoes of the same style consecutively. Small batches are the norm and it is very rare that shoes of identical size and style are processed through the factory together.

It is likely that shoe manufacturers will be asked in the future to process smaller and smaller batches of shoes and to respond to retail requirements for repeat orders more quickly. This means that flexible automation of small batch production will be one of the keys to the application of new technology.

Shoe manufacture may be conveniently split into two phases. The first covers the processing of upper components in the flat, limp state through cutting, preparation and joining operations (sticking and stitching) where the shoe begins to take shape. In the second phase the shoe is in the 3 dimensional form, starting with the completed upper, fitting and wrapping it around the last, then processing this through to sole attachment. The two phases will be considered separately looking at the material and handling problems, particularly for shoes with stuck-on soles.

Upper preparation

Leather remains an important material for shoe uppers and both comfort and fashion dictate that this will continue for the foreseeable future. Leather varies in type and physical properties, skins used for uppers can be between 0.4 mm (goat) up to over 3 mm (cow) and these naturally give a large variation in drape and handling properties. There has been a trend over the past few years towards softer, more flexible leathers. The main problem with leather is that it comes in irregular shaped and sized skins and both quality and characteristics are variable within a skin as well as between skins. Marks or faults in the skin have to be identified and avoided when cutting. Leather is non-homogenous in its physical properties meaning that specific components can only be cut from certain parts and in a particular orientation. All this means that cutting upper components from leather is a highly skilled operation.

Synthetic upper materials offer fewer problems in cutting due to their consistency and regular sized sheets or rolls (however, their inferior stretch and compression properties require more care in manufacture). A wide range of materials is used, mainly in women's footwear, including fabrics and coated fabrics (woven, knitted and non-woven) as well as some homogeneous polymeric materials. In general, synthetic upper materials cover the same wide range of handling properties as leather.

Most upper components are cut from sheets or skins using hand made press knives on manually controlled cutting presses, with the knife being re-positioned manually on the material between each stroke of the press. The cut pieces are picked out by hand and sorted manually. For larger production runs of some synthetic materials, NC cutting presses are available but not in widespread use. There is currently considerable interest in the use of water jet cutters for both leather and synthetics, the cutting lay plan being generated on a computer beforehand. However, with all computer controlled cutting at present, the cut components have to be picked out, sorted and marked by hand. It is understood that some research is being undertaken in the use of a robot for picking out cut components.

The flat, cut components are normally grouped in 12 pair lots so that all pieces for one style are in a box. This passes through the upper preparation area from one operation to the next, either by hand or with the aid of a conveyor or mobile rack system. Initial operations involve marking, reducing thickness of edges and folding, adding decorative cut-outs and reinforcements and these are followed by the stitching (or closing) operations. In these the various components are sewn together in stages. By far the majority of operations require manual handling and positioning of components under the machine head.

Computer controlled stitching machines are increasingly used in some factories but are limited to decorative stitching or simple constructional seams where quality does not need to be high. Components are precisely located in pallets, which are then moved under the stitching head on an X-Y table using NC data. Pallets have to be loaded by hand however and introduced to and removed from the stitching machine manually. By far the great majority of stitching is still carried out on manual sewing machines.

The upper preparation and closing processes involve the greatest use of labour in a shoe factory. In an average factory employing 150 operatives, half will be in the upper preparation and stitching areas. Furthermore, the time which a shoe upper takes to pass through the closing room is large. There is thus considerable scope for the use of new technology in both reducing labour costs and reducing throughput time.

However, there are a number of problems. One of these is that different styles of shoe require different operations and sequences of operations and it is impossible to create a flow line working arrangement. Work may need to go backwards and forwards through the closing room in a seemingly haphazard manner. A further problem is the variation in number, size and shape of the complex upper components which make up a shoe. Handling and manipulation of these presents a real problem.

There is thus a need for systems which can identify, handle and process single components and groups of components of varying size, shape and flexibility, systems which can feed and control a wide variety of machines for marking, selectively reducing thickness, folding and stitching and to route components to the correct machine according to the style. Relatively little work has so far been undertaken in this area and no solutions are being offered commercially. It is perhaps a challenge to the use of sensory robotics.

Handling the shoe

Lasting is the process whereby the stitched upper takes on its 3D shape by stretching and wrapping it around the plastic former or last, which approximates to the shape of the foot. Lasting machines are highly complex mechanically but still require human manipulation and skill to locate the flexible upper in the machine correctly and to judge the degree of stretch required to suit the material properties.

All subsequent operations are carried out on the "lasted upper". These normally include roughing the bottom surface or lasted margin, applying adhesive to it, then drying the adhesive. The pre-fabricated sole is introduced at this point with adhesive already applied. Heat is applied to activate the adhesive. This will often cause the sole to become highly flexible and difficult to handle. An operative then "spots" the sole in the correct position on the shoe bottom at toe and heel. The adhesive being tacky, the operator has to be correct first time, no sliding is possible. The shoe is then put onto a sole press to complete the bonding process. Last removal is the final making operation.

Lasted shoes are normally passed by hand between operations or moved around the production line on conveyors or mobile racks, in which case they are passed manually between conveyor or rack and machine. Often a hole in the top of each last fits onto a peg on the rack or conveyor. However, relatively few processing operations utilise this hole to locate the last in the machine, and consequently its consistency in position and size is very poor.

Some automation has been introduced into the lasting and making area. Automatic roughing machines for example use information which is pre-digitised. The shoe is held at the toe and heel by grippers which sense the shoe size and left or right foot. This automatically adjusts the digitised data to the correct size and foot.

Robots are being used to rough around the edge of shoes using CAD data to define the shape, but with pressure sensing to adjust the roughing tool to variations in upper thickness at seams etc. They are not designed to cope with significant changes in shoe style, but are for long production runs. They use a special location plate accurately fitted to the top of each last to position the shoe precisely.

There would appear to be fewer problems in automating lasting and making activities in the shoe factory compared with the upper preparation activities. This is mainly because the upper is constrained to the shape of the last. A combination of JIT techniques of re-organising the grouping of lasting and making operations, coupled with the use of some automated machinery, mean that only 4-5 operatives are required to operate a lasting and making line, compared with 10-15 a few years ago. Further progress is likely to be hampered by the lack of automated quality control procedures, which mean that human inspection is needed to maintain quality and consistency.

SATRA's sole attaching robot work cell

SATRA has for a number of years been experimenting with the use of robotics in shoe manufacture. The aim has been to ascertain the problems of utilising such technology in a shoe factory and to determine if a flexible work cell can be developed which will form a viable economic unit.

The sole attaching task was chosen as a group of operations which utilised more than one operative, and suffered from problems of lack of quality and consistency. The objective was to develop a demonstration work cell which would apply adhesive to the shoe bottom and "spot" the sole onto the upper ready for pressing. The project also aimed to determine the usefulness of 3D CAD data to define the position and co-ordinates of the shoe upper, as well as vision sensing to detect the position of shoe and sole at various stages in the cell cycle. The vision system used was based on one devised by Hull University and they were involved in the initial development of the sole spotting process which is described in a separate paper.

The work cell uses a PUMA 560 robot running under VAL II. The cell (figure 1) will accurately attach the correct sole to any lasted upper, using a gripper mounted camera to locate the last and sole as well as a CAD description of each last. A cell control computer orchestrates a bar code reader, the vision system, the robot, CAD database and ancillary equipment, communicating through simple serial lines.

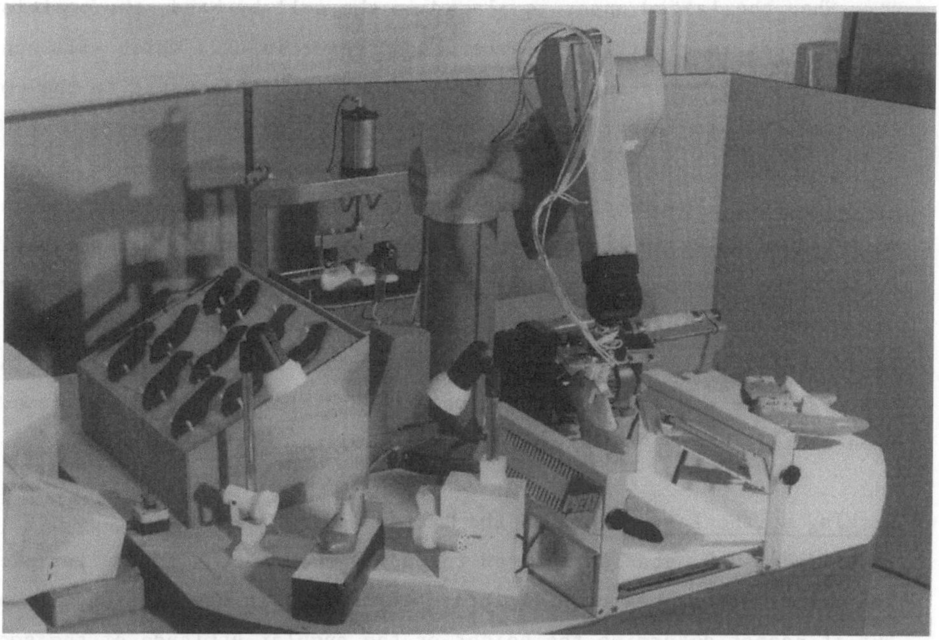

Figure 1 - Sole attaching work cell

The work cell demonstrates the following steps :-

- Identifying a last using a bar code attached to it
- Picking up the lasted upper
- Applying adhesive to the lasted upper
- Placing the lasted upper in a dryer
- Selecting the sole and placing it in the "spotting" rig
- Heat activating the sole
- Collecting a lasted upper from the dryer
- Spotting the lasted upper onto the sole using the vision system to locate the sole
- Pressing the assembled shoe in a bonding press.

<u>Identification of the last</u>. The last must be identified as soon as the shoe enters the cell so that the correct data can be used for picking it

up. When the lasted upper is placed in the cell by hand, the position of the bar code is uncertain so a bar code reader is used which will hunt for a bar code, identify and decode it and then send the result to the control computer, all in less than a second.

End effector. The robot end effector (figure 2) has a suction cup which will handle soles in their less flexible state prior to heat activation, a solid state camera and a gripper for the lasts. Lasts are gripped by a pin which, once inserted into the socket hole in the top of the last, can be expanded. A smaller front pin ensures that the last is held at the correct angle and micro-switches detect when the pins are fully home. Any type of last can be held providing it is fitted with a socket.

Picking up the last. Because the lasted upper has already been identified, CAD data can be used to modify the robot's motion for pick up. The robot mounted camera will locate the socket in the horizontal plane but it cannot determine the height or the angular attitude of the socket which varies with the style and size of the last. A computation using the CAD description of the last bottom gives these values rapidly.

Cementing the lasted upper. The CAD description of the edge of the last bottom is converted to a set of robot coordinates so that the robot can move the upper over a conventional roller sole cementer. The flexible roller accommodates the variations in thickness of the roughed upper.

Drying rack holds last. A prototype drying rack can hold any last by making use of two holes in the side of the last near the socket. The robot can slide the last onto horizontal pins from the side, release it and then withdraw vertically. The last can be picked up by the reverse process.

Sole selection, activation and spotting. Soles are supplied ready cemented, as in the factory. They are stiff enough to be handled as rigid components in their unheated state. The robot uses its suction cup to select the correct sole for the last and place it on the "spotting" rig where the purpose built heat activator can be triggered by the robot computer. By heat activating the adhesive on the sole in the spotting rig, the problem of handling sticky, flexible soles is completely avoided.

Also, when it is on stand-by, the light from the activator serves to illuminate the sole for the robot mounted camera as it determines the exact position of the sole before bringing the lasted upper down onto it.

To force the sole into the waist of the last a bladder in the rig base is inflated (figure 3). Higher heeled shoes need more inflation and this is determined from the CAD data.

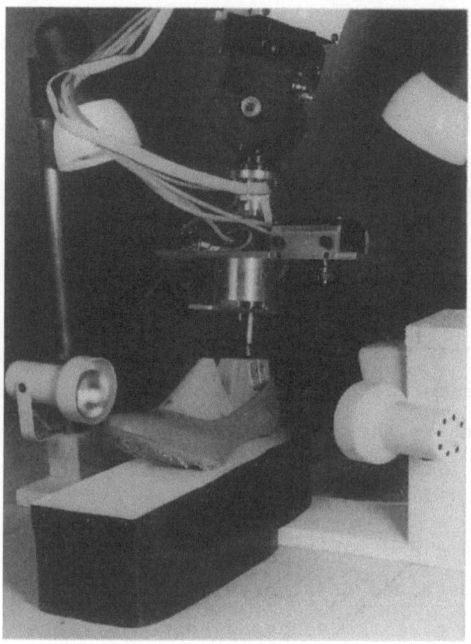

Figure 2
End effector picking up shoe

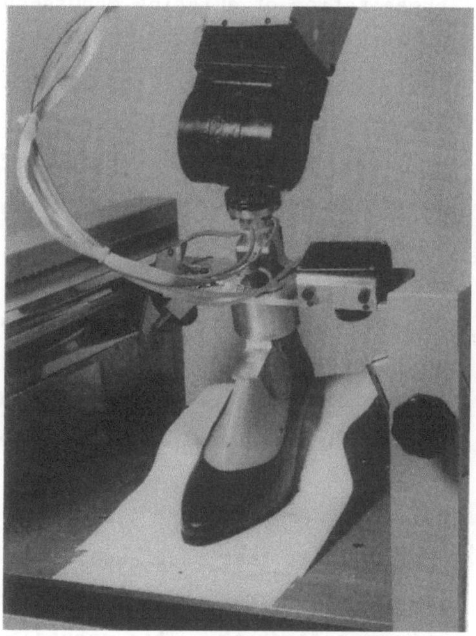

Figure 3
Sole spotting

<u>Bonding press accommodates all lasts</u>. The robot deposits the lasted shoe onto the bed of the press and is then free to start on the next cycle while the bed and shoe slide under the press heads which descend under low pressure to push the shoe into the base. The front of the base rises to reach the toe of high heeled shoe lasts, using their CAD description as a

guide, before the flexible base is 'frozen' and the full bonding pressure is applied for about ten seconds. The completed shoe is then ejected from the cell by inflating the base, allowing the shoe to be swept off.

This demonstration cell shows that it is feasible to attach soles automatically. However, there are many problems to be overcome. The method of last handling was dictated by the low load handling capabilities of the PUMA 560. Utilising the hole in the last would not involve extra tooling costs but would involve more sophisticated sensing and detection to locate precisely. A gripper plate could be designed which would require less accurate sensing but would require considerable re-organisation of existing manufacturing techniques.

The use of 3D CAD data to define the shape of the shoe can in theory avoid the need for sensing, but in practice the variation between similar lasts and the variations in thickness of the uppers and components mean the information is not fully reliable. Also, the Puma used gave path errors when using external data.

Experience with vision and other sensory systems suggests that the speed of processing data may be critical. There was some evidence to suggest that a more sensitive system would be required to give the necessary accuracy. This would inevitably exacerbate the problem of processing speed.

It is likely that most successful sensing and following systems will need to use a combination of both external data and sensing which will act as adaptive feedback. This should help to give the necessary accuracy with the required speed of operation.

Conclusions

To maintain a viable footwear industry in high labour cost countries, it is essential that methods of reducing cost of manufacture and speeding up response to retail demands are evolved. The use of new technology will be vital in achieving that aim. Flexible automation will be one of the key roles in coping with the wide variation in sizes and styles of footwear which factories need to produce.

Most effort in the past few years has been concentrated on automation and reducing labour in the lasting and making areas of shoe production. Whilst many of the concepts of last handling and robotics and sensory applications have still to be fully exploited, the benefits to shoemakers of more fundamental research in this area will be limited.

Major benefits await shoe manufacturers when systems are available which reduce labour costs and reduce throughput time in the upper making area. The problems of handling of limp materials, processing and stitching them together using automated techniques are immense, but it is anticipated that sensory robotics will play an important role in the development of new systems for upper preparation.

It is suggested that the use of CAD data coupled with sensory feedback may be required to obtain a practical combination which would meet the demands of accuracy and speed of automated processes, taking into account the variations in style, size and quality of materials and components which are used in the shoe industry.

AUTOMATED APPLICATION OF SOLES TO SHOES

I. Gibson, G.E. Taylor and P.M. Taylor
Department of Electronic Engineering
University of Hull
Hull HU6 7RX, U.K.

Keywords: Shoe assembly, robotics, automation, low resolution vision, CAD.

Abstract

Robotic assembly of shoes is discussed with a specific example of application of soles on to lasted shoe uppers. A master/slave construction links a Puma 560 robot to sensory information from low resolution cameras. Comment is also made concerning the significance of CAD information to the manufacturing process.

1. Introduction

The project to be described here started in 1984 with the aim of contributing to ongoing work by SATRA designed to automate an entire section of a shoe assembly line, using robot accuracy and consistency to improve the quality and reliability of the end product. The particular part of the process considered at Hull was automation of the shoe sole attachment process.

The main difficulty which occurs in aligning and bonding soles onto lasted uppers occurs when making the initial contact between the two parts. This contact, known as spotting, requires the use of considerable visual and tactile information when carried out manually since toe and heel areas have to be manipulated in turn to obtain a close fit. Sufficient pressure must be applied to ensure the sole remains in position while it is transferred to a mechanical press which produces the final bond.

The paper will briefly review early attempts at automation before describing in detail the final system which was implemented as part of a complete work cell demonstration for SATRA members.

NATO ASI Series, Vol. F 64
Sensory Robotics for the Handling of Limp Materials
Edited by P. M. Taylor

2. Initial Work

In initial experiments the robot was used to manipulate the sole unit above a fixed and inverted lasted upper. This strategy was chosen because, at that time, grippers available to handle the upper were such that the resulting payload was too heavy to allow sufficiently accurate manipulation by the PUMA robot. The original prototype workstation used two fixed cameras placed above the critical toe and heel areas of the inverted upper. The PUMA 560 robot was used to bring the sole into approximate position above the upper (and thus into view of the cameras). The sole position was then adjusted to minimise the non-background area measured by the cameras, the minimum occurring when upper and sole were perfectly aligned. The main difficulty with this approach was that to obtain sufficient accuracy in the alignment it was necessary to have the sole/upper fill virtually all the area of the image making it difficult for the system to cope with left and right shoes of varying sizes and styles.

An alternative strategy obtained information about sole and upper separately via a gripper mounted camera. This information was then used to allow joining via a simple open loop pick and place operation. The difficulty with this was that the cameras gave information only about a silhouette of the three dimensional upper. This is not a true representation of the surface to be bonded and it was found impossible to achieve the required accuracy of fit.

Discussions with SATRA suggested that a solution to this problem might be found by linking CAD systems now being used increasingly in shoe design to the robot workstation. This enables the robot to obtain a profile of the bonding surface in the form of a stream of locations relative to one or more specific points. The sensory information required from the robot is then reduced simply to locating this point. As part of this approach a new and much lighter gripper was designed to handle the uppers allowing the relative position of the components to be reversed. Sensory information is again obtained from two cameras located above the assembly area and using this together with the CAD information the upper is adjusted until it is aligned with the toe part of the sole. A spot contact is then made to begin the bonding process which is completed by inflating a cushion under the front part of the sole unit until the sole is firmly positioned under the upper (see figure 1). The robot then moves the heel area into place and transfers the whole assembly to the sole press. A detailed description of the complete process is given in the next section.

3.Detailed System Description

This is the system shown in Figure 2 and used for the final demonstration module and is the same as that demonstrated at SATRA for members day, 12th May 1987. It was demonstrated in conjunction with a glueing station developed at SATRA [1] and adapted for control by the same central control unit. This was seen as a demonstration of the versatility of the system, requiring no further electronic hardware not essential to the gluing operation alone.

3.1 System construction

The diagram in Figure 3 shows the structure of the assembly control system. The system is based on the master-slave principle where the master is designed to control the communications and coordinate the actions of the slave devices. A typical slave device would be a robot, or any other sensor/actuator combination (cameras, X-Y tables etc.). The master controller is based on the LASCAR system [2]. This system was developed within the department as a research project and was intended to be a complete operating system for robot assembly. High level commands could be constructed to perform specific assembly tasks, for example:-

(1) GET AREA OF PICTURE;

(2) SEND AREA TO ROBOT;

would both be valid LASCAR commands. Taking example (1), PICTURE is interpreted as a device, in this case probably an image processor, and from this LASCAR would know which slave to talk to. GET would be a command to request an input from the image processor and the input, which hopefully is an area value which can be put into the variable AREA. The fill in word OF is an option used in the device command builder to make the command line more readable. The second example illustrates how the LASCAR system would cope with the transmission of data to a slave device using SEND as the command, AREA as the parameter, and ROBOT as the destination. Note that although these two examples appear similar, the underlying data format may be significantly different. For example, the first command may be sent as an ASCII string in the first example, but as a binary sequence in the second.

Slaves communicate with the LASCAR master via a common bus termed ROBUS (RObots BUS [3]). This is a parallel bus designed as a general purpose connection between robots and peripherals. All slaves connected to ROBUS have a standard interface with 8 address lines, 8 interrupt lines, 8 data lines, plus a few handshaking lines. Each device has an address that uniquely defines it to the master. The type of system is dependent on the user and the purpose, it can be interrupt driven or polled. A standard ROBUS interface [4] has been designed using a Z80 microprocessor. The purpose behind this is to ease the hardware design burden on a system designer and provide him with the basis for a 'smart' ROBUS slave. The slave card is very versatile, having a direct memory connection to ROBUS, comprising

1. Twin serial i/o ports with software programmable baud rates.
2. Intel 8255 parallel i/o port providing three, 8 bit ports.
3. Software programmable timer.
4. Four indicator LEDs and an audible warning device.
5. Two jumper selectable 8k byte 'Jedec' sites for a variety of on board EPROM and RAM chips.
6. Fully implemented interrupt structure using the Z80's mode 2.

Item 5 means that it is possible to program the slave card using a number of EPROM types. This has recently been modified to cope with any normal EPROM size up to 128k. The software can be written in straight Z80 assembler, or via a Pascal/Z80 cross-compiler. A lot of the software (standard communications routines, hardware initialisation, etc.) has already been written and put into libraries for ease of access.

3.2 Robot

The robot used throughout the project was a Puma 560. For the experiments on inverted assembly this robot was controlled using the VAL [5] operating system. The fairly rudimentary nature of the VAL operating system is not a burden since it is required to do fairly menial tasks. The more intelligent operations are performed by the LASCAR master.

The choice of the Puma 560 was a matter of convenience rather than necessity. The fact that software would have to be compatible with that written at SATRA for future amalgamation of systems was important since SATRA also possess a VAL2 Puma. A six degree of freedom robot is very useful as a robotic research tool. The number of degrees of freedom anticipated is usually more than ultimately used. When developing a robot based system it is useful not to have to worry about whether it is possible to reach a particular point in space. The final system would have an optimal number of degrees of freedom to reduce robot cost. It was anticipated that a rocking motion may be required for the shoe sole assembly process and so at least five degrees of freedom (i.e a robot 'wrist') were deemed necessary for the initial experiments.

Control of the Puma was through a general purpose slave card. One of the serial lines was connected to the console terminal port of the Unimation controller for the robot. It would be possible to use the supervisor port on the controller unit which would enable the user to run programs from the console terminal as well as externally. This supervisor option is only available with a very few expensive robot controllers, and they differ from system to system. The DEC protocol (DDCMP) used by the Unimation controller is available only to Digital machines and is very difficult to synthesize using the general purpose slave as a terminal input device. Each individual operation is written as a separate program in VAL2. These programs are executed from the LASCAR system. In this way, operations can be modularised and updated very quickly. Individual signals directly associated with robot motion are interfaced directly to the robot signal i/o lines. These include signals to switch on air pressure and micro switch input lines. These were used because of their availability in the VAL2 system. Should a robot system be used in the future that does not have the facility of i/o lines, the parallel ports on the general purpose slave card could be used instead.

3.3 Cameras

The cameras used in the final system were a commercial version of the DRAM camera, similar to those developed at Hull[6]. The EV1 'Snap' camera, produced by Micro-Robotics of Cambridge is based around a BBC micro computer [7]. The camera uses both sides of the IS32 memory chip and is intended as a medium resolution device (128x256 pixels).

External control of the BBC is enabled via the RS423 serial interface. This is connected again through a serial line on one of the general purpose slave cards. Part of the 6502 assembly routine for the camera sets up a parameter block containing various details concerning objects in view. Area, perimeter, centre of area, position, and dimensions of the enclosing box can all be extracted from this parameter block. The main program running on the BBC continually updates the image data and scans the input buffer for a command from the LASCAR master. LASCAR requests a parameter for the image and the BBC responds with the data in a fixed format. The speed at which the camera responds depends on the lighting level. The vision system reads the IS32 at a rate of 1MHz. This means that the fastest a frame can be read into the BBC is 60ms. This corresponds to quite a high lighting level and a more likely rate is 0.3s/frame.

LASCAR has a FORK facility, enabling processes to run in parallel so long as separate slave devices are accessed in each branch of the fork. This is very useful for occasions where a robot is required to do a separate task from the vision system. In this case, for example, it is not required that a vision system operation is completed before a robot task.

3.4 Gripper construction and pick up

The gripper is designed to pick up lasted uppers. It is assumed that the lasted uppers will be presented in an upright form with the locating pin hole visible to a camera mounted on the gripper. The gripping technique used is an internal jaw technique, with forces exerted by the external sides of the jaws. One of the jaws is in the form of a pin, fitting into the locating pin hole. The other jaw is a plate, designed to push against the flat at the front of the last (Figure 4).

Lamps are placed and a threshold selected in such a way that the heel part of the upper is indistinct from the background. Since the hole is quite deep, the light does not shine down into it and so appears as the darkest object in the field of view. The centre of area of this hole image is then used to locate the upper. The hole image is brought into the centre of the field of view. Once this is done the position of the hole can be described relative to the robot position. The locating pin on the gripper is then inserted into this hole.

The top of the last with the locating pin hole does not lie parallel with the horizontal (X-Y) plane of the robot. The hole is, however, drilled perpendicular to the top of the last. Since the locating pin fits quite tightly into the hole, the angle of approach is fairly critical since the pin will catch in the hole if inserted at the wrong angle. This angle of approach can be determined from the CAD profile of the lasted upper.

3.4.1 CAD data

The CAD data is stored as a series of points relative to a datum or reference point. The reference point that is considered the most logical is the centre of the locating pin hole. Although the entire net of the lasted upper can be digitised using a CAD system, only the sole area (the bonding profile) was used in this experiment. This data was simulated CAD data and not actually generated from an existing system.

3.4.2 Pick up

The angle of approach for pick up can be estimated from the CAD data. When the last is left to rest in an upright position, it can be seen to touch a flat surface at two points. One point of contact is at the heel area, the other is at the opposite end of the waist where the ball of the foot is. By taking the point at the rear of the heel, the first point of CAD data, one point can be extracted for the angle of approach (point A). The other point is taken where the height variation from the reference point is greatest (point B). These two points form the hypotenuse of a right angled triangle. The angle of approach Θ is defined by:-

$$\Theta = \tan^{-1}\left[\frac{(A_Z - B_Z)}{(A_X - B_X)}\right]$$

The actual sign of the angle is not important since this will depend on the attitude of the initial CAD data. The angle of tilt will always be back towards the heel since there are no shoes that tilt the other way. This angle of tilt can easily be programmed into a robot as a rotation about the tip of the locating pin on the gripper. Calculation of the angle can be performed by the robot controller (if it is capable of trigonometric functions) or by the LASCAR system.

This angle of approach is of course only an approximation taken from the original data of a disassembled last. Slight variations, in the z-direction, in particular, can result from excess sole filler or strengtheners incorrectly placed in the waist of the last. The angle of approach can be determined sufficiently to avoid snagging of the pin during insertion, providing a compliant undersurface is used to take into account any slight variations. From a sample of 8 lasted uppers, all the same style, a variation of $<0.5^\circ$ was recorded.

3.4.3 Last design

Most lasts produced today are not produced using CAD data or with robots in mind. As a result they are produced with convenience and speed in mind without much attention being paid to accurate finishing off. The reason behind this is not because it is impossible (or uneconomical) to do this but because shoe manufacturers do not require lasts to be made in this fashion.

If robots are to be used in conjunction with CAD data, then it is necessary that greater care is taken in the last design. With the centre of the locating hole being used as a reference point for the CAD data, this hole must be drilled with accurate jigging to maintain the relative locations for the last. By having a fixed hole location the last can be correctly positioned within a robot gripper. There is still however a problem concerned with the rotational error involved during the pickup process.

With the hole being circular, the last can be positioned in the gripper in a number of orientations. The orientation cannot be determined from the fact that the gripping plate normalizes the forces to a fixed orientation but this factor can be used to orientate the last coarsely. By taking the line from the toe point and the centre of the locating hole and drilling a second hole with its centre on this line, a gripper can be built to utilize this three point contact to uniquely define the last to the upper [fig 5.4]. Since the third pin that is inserted into this hole is used only for location and not gripping, a single direction compliance has been built into the design. A chamfer on the pin allows the gripper to bring the last into correct orientation if there is a slight error. Tests have shown that a last can be positioned with a disorientation of +/−5° from that of the gripper and still bring it into alignment during the pick up sequence.

3.5 Spotting operation

The sole spotting operation involves correlating between data obtained from cameras concerning the sole units and data obtained from the pseudo CAD system describing the upper. Since the upper is theoretically clamped in the gripper in a fixed position and orientation, any error involved will result from the location of the sole. The errors will result in positional and rotational misalignments. The sole will be positioned in the assembly area with reasonable tolerances. A feeding mechanism can be easily designed that will position the sole for assembly within +/−40mm of correct alignment at both the toe and heel areas. For a sole 210mm long this will result in a rotational error of approximately 20° one extreme to the other.

An EV1 'Snap' camera, fixed above the toe area of the sole unit provides a point that can be correlated with that of the lasted upper. All the points from the CAD data are known to the robot, this toe point is particularly useful since it is aligned with the robot TOOL axis. The transfer of the robot tool to this variable location from the fixed pin location is therefore a simple translation along a single axis. The correlation procedure relies on a calibration point known to both the robot and the camera. A shift along the axes of the camera of the error value between this calibration point and the given tip location results in the actual tip on the last lying above that of the sole.

By defining the tip point as a tool location it is rotated about this tool centre, maintaining its position. A camera placed above the rear of the sole unit is used to calculate the rotational error resulting from incorrect positioning of the sole unit. The view from this camera is a parallelogram. If the locating hole on the last lies centrally in the sole area (which it does not), then the centre of this parallelogram would lie on the same line as that between the tip and the pin hole. A simple piece of geometry resulting from a calculation between the distance from the front camera window and the rear and the error data from the front and rear camera views gives the angular offset. Also, a small adjustment is required to take into account the offset of the locating pin hole from the centre of the heel area. This can be achieved by looking at the CAD data for the points adjacent to that of the pin (0 y-offset). By taking the arctan of the y components of the various pieces of data and dividing them by the x components, the resulting angle can be transferred to the robot for it to modify its position.

The positional and rotational corrections are conducted before any contact is made between the upper and the sole. Since the full operation will involve a sole unit with activated adhesive on it, any contact will result in adhesion. With the sole not being held in position in any way, all the camera data will be lost and irretrievable. The final obstacle to overcome in this sole spotting process involves the contoured nature of the uppers with respect to the relatively flat surface of the sole units. The problem lies in how to mould the sole unit around the upper.

The robot is used to clamp the upper to the sole at the tip location. This is achieved by approaching the point at an angle. Rather than bring the upper down on the sole, this position is maintained by the robot and the compliant undersurface the sole is lying on is moved upwards, pushing the sole into the upper (see figure 5.5). The movement is achieved by inflating a bag underneath the compliant (foam rubber) surface. Since the inflation of a bag is an equalising force and the distance travelled is greater the further away from the tip, the sole is pushed back from the tip to the heel. This avoids any buckling effect, and if used in conjunction with robot motion may be used to enable the stretch effect required for undersize sole units.

4. System Performance and Future Work

Tests performed during the SATRA member's day showed that the system was capable of spot connecting lasted uppers to sole units with some degree of success. Of a total of 100 monitored trials carried out both at SATRA and at Hull, only 8% resulted in wholly fatal errors, requiring manual intervention to reset the system. 65% of the runs performed adequate sole spotting with the required accuracy. All the errors were traceable with the majority of them rectifiable. Such figures, although far from perfect, are sufficient to confirm that the system concept works and further development is worth pursuing, using these ideas, to produce a functioning robotic assembly station. The system has now been transferred to SATRA and work continues there.

References

1. Hackney B., Potential of Robots in Shoemaking, SATRA Bulletin, Nov. 1986, p 255.
2. Taylor P.M., Stubbings C.A., LASCAR: A Language for Assembly Station Control at Runtime, Proc. 4th IASTED Symp. on Robotics and Automation, Amsterdam, 1984, pp 86–89.
3. Taylor P.M., Stubbings C.A., Software and Hardware Aspects of a Flexible Workstation for Assembly Tasks Using Sensory Robotics, Proc. 2nd IASTED Symp. on Robotics and Automation, Lugano 1983, pp 48–51.
4. Stubbings C.A., General Purpose ROBUS Z80 Processor Card, Hull University Dept. Electronic Eng., Sept. 1984.
5. Users Guide to VAL Version 12, A Robot Programming and Control System, Unimation Inc., 1980.
6. Taylor G.E., Taylor P.M., Optimal Use of Low Resolution Vision Sensors, Proc. IASTED Symp. on Modelling, Identification and Control, Grindelwald, Switzerland, Feb 1987, pp 219–222.
7. Van Sommen A., The Snap Camera Manual, Micro Robotics, Cambridge, 1984.

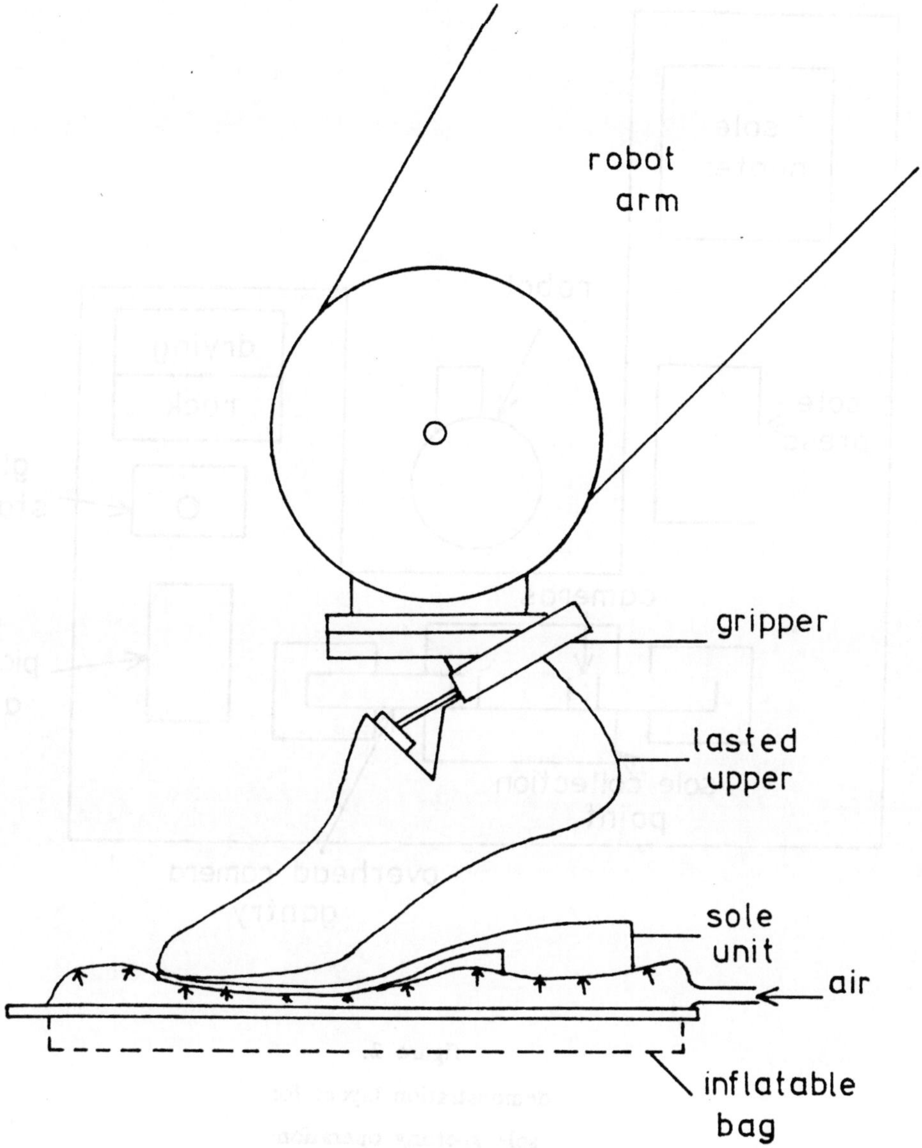

figure 1.
principle of compliant bag used
as spotting surface

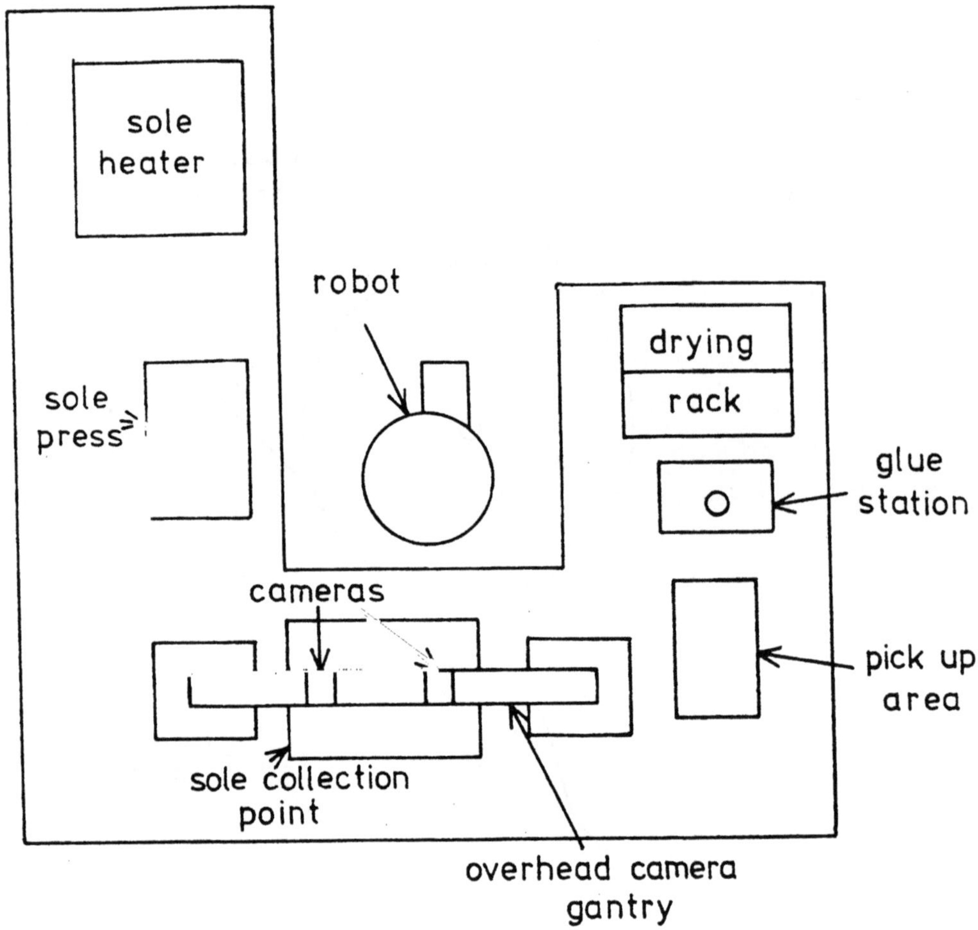

figure 2.
demonstration layout for
sole spotting operation

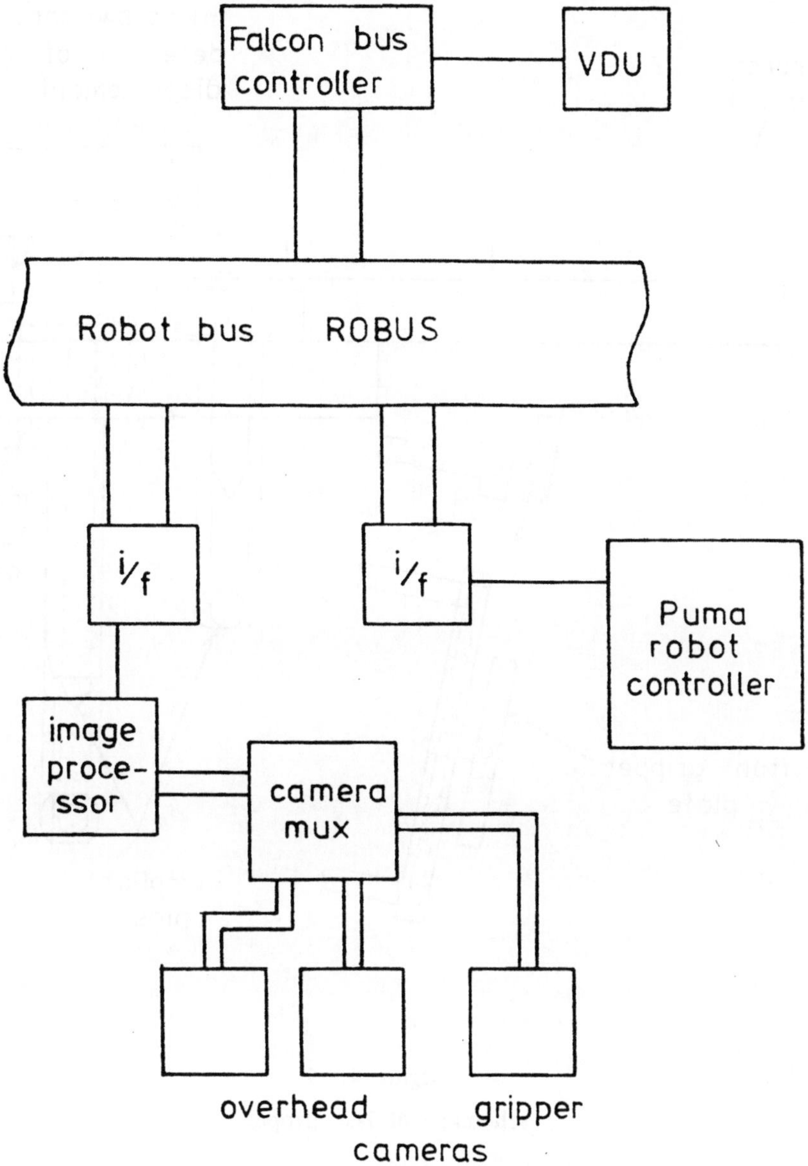

figure 3.

schematic of robot control system for sole spotting demonstration

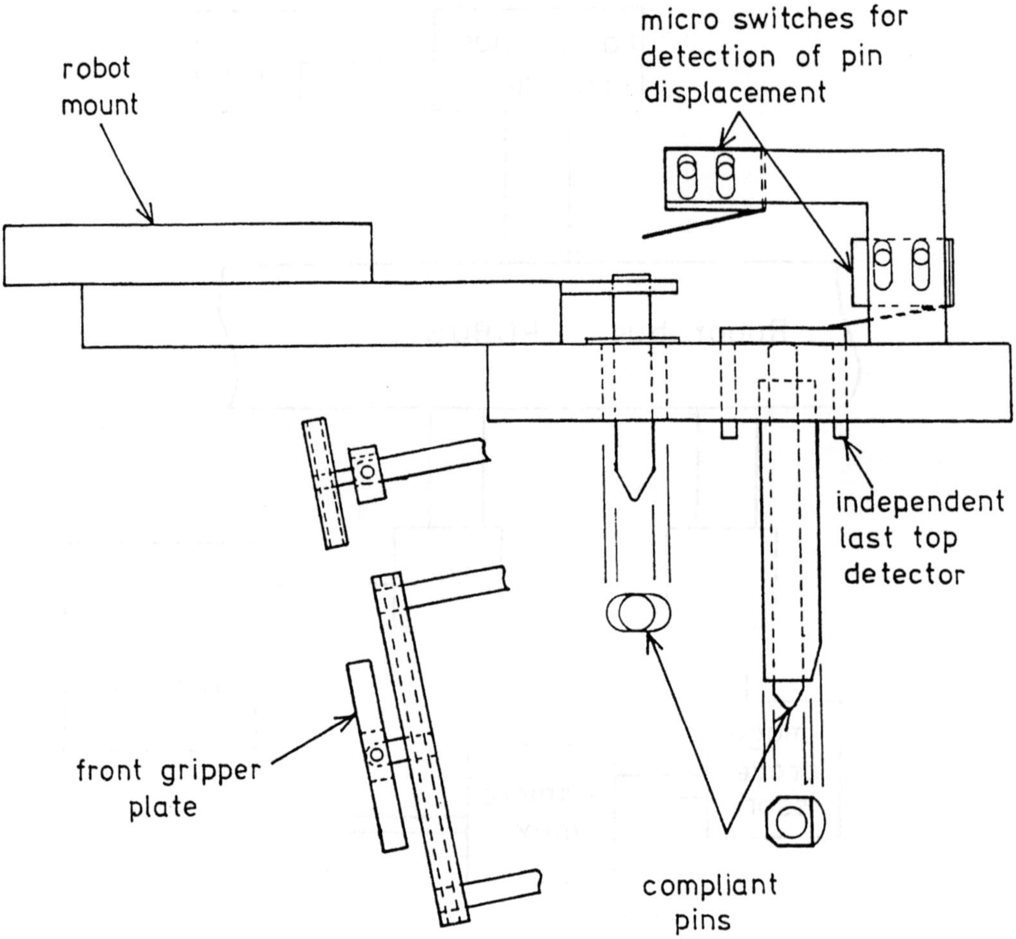

figure 4.
schematic of last gripper

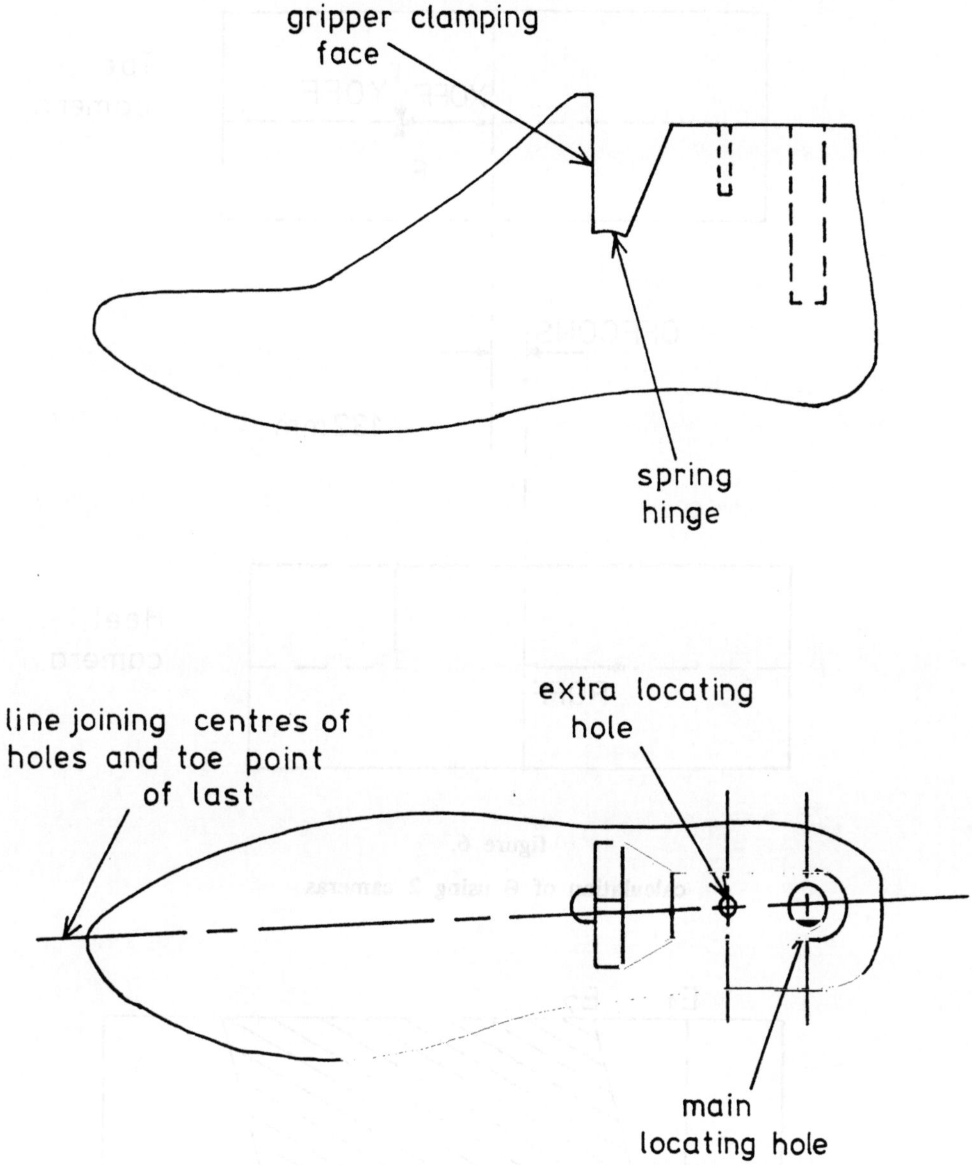

figure 5.

augmented last design for

robot handling

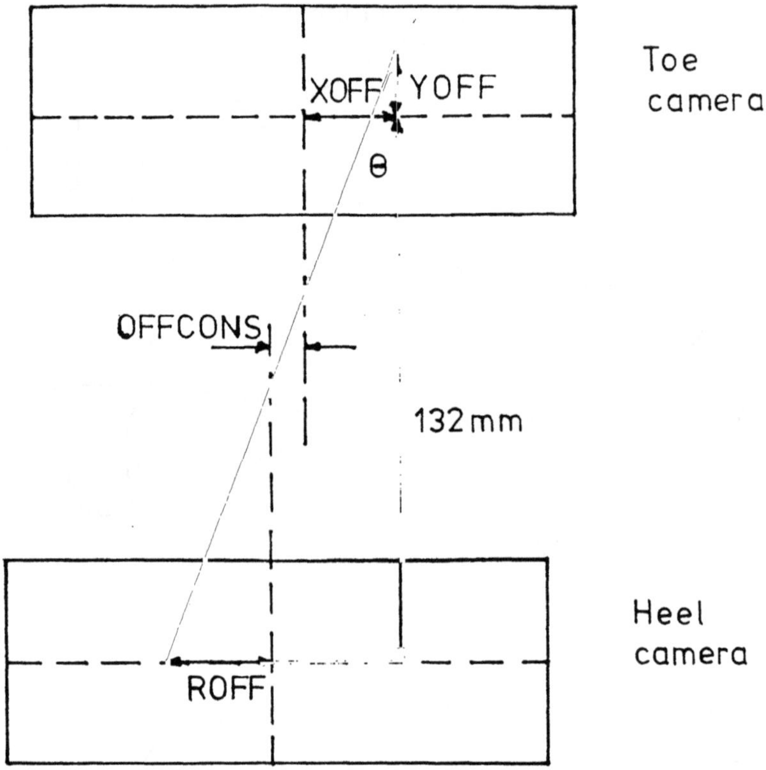

figure 6.
calculation of θ using 2 cameras

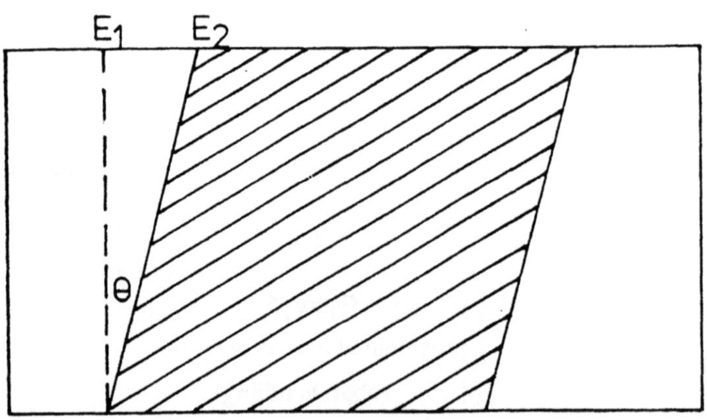

figure 7.
Calculation of θ using
data from heel camera only

Round Table Discussion
Sensory Robotics for Shoes: Trends and Future Needs

Chairman: Dr. J. Simmons
Secretary: Mr. G. Monkman

The round table discussion started with the statement made by Mr. D. Reedman of BUSM ltd that "*Robots are presently inadequate for shoe assembly*". It was suggested that the mechanics was the problem, not the electrics. The final concensus seemed to be that a systems approach (as a whole) should be pursued, rather than replacing industrial operations with single robots.

The conversation then turned to the material being handled rather than the handling devices. The use of synthetics was discussed. Most people concerned with the shoe industry agreed that synthetic materials were better but that they did not sell! From an engineer's point of view, it appeared that more information on the mechanical properties of leather would be desirable.

Returning to robotics, Dr. Gaynor Taylor of Hull University stated the use of error recovery and sensor implementation and fusion as a way forward. It appears that most of the sensors used in robotics for leather and textile handling are a result of research in other areas. No tactile sensors are presently commercially available.

The importance of maintaining uniform force on leather during processing was introduced and the discussion resulted in the talk of chemical adhesives (some needing no pressure), and the question of who makes presses. No presses for this purpose are available yet.

The cost of robots was generally considered too high, and their speed too slow. Though costs are descending, and as pointed out by Prof. Bob Kelly (Renesslaer Polytechnic) most of the time is wasted having components in motion. A general discussion of whether robots should or should not be anthropormorphic ensued with no real consensus being acheived on this point.

As a particular requirement, Mr. David Reedman stated a need for a camera with a resolution of 100 microns, capable of at least four frames per second at a cost of less than $5000. This sort of high grade, low cost device appeared to be generally most desirable, but unavailable.

NATO ASI Series, Vol. F 64
Sensory Robotics for the Handling of Limp Materials
Edited by P. M. Taylor

ROBOTIC HANDLING AND LAY-UP OF ADVANCED COMPOSITE MATERIALS - AN OVERVIEW

M. Sarhadi
School of Engineering and Applied Science
University of Durham
South Road
Durham, DH1 3LE
U.K.

NATO ASI Series, Vol. F 64
Sensory Robotics for the Handling of Limp Materia
Edited by P. M. Taylor

1. INTRODUCTION

In recent years, the aerospace, automotive and other industries have achieved significant product innovations by substituting metal materials with advanced composites. High performance, increasing requirements for work precision, productivity of the manufacturing process and keen industrial competition are strong incentives for continued growth in the use of these new materials. Composites already account for half the structural weight of some advanced tactical fighter aircraft and it is expected that by the turn of the century roughly the same proportion of the structural weight of commercial aircraft will be made up of composites [1]. The new Boeing 360 commercial helicopter is made almost entirely of composites and in Boeing's next generation of planes up to twenty percent of the structure will be made up of composites. There is also a growing interest in the development of structural applications of these materials in the automotive industry. A recent report predicts that provided significant advances in manufacturing and assembly technologies are made, and the design capabilities are improved, the use of advanced composites by the automotive industry will increase dramatically during the 1990's [2]. The attraction of advanced composites in the motor industry will be lower manufacturing costs than traditional steel construction and faster, more frequent model changes, says the report.

Unfortunately, the processing of advanced composite materials requires meticulous craftsmanship and is very labour intensive. This greatly increases the production costs, and often results in manufacturing defects [3]. It is very important to ensure correct and accurate fibre alignment within the component if high-performance properties are to be achieved. It is also necessary to maintain the fibre content and distribution accurately. Controlling these factors will give a component of consistent quality. Automation of the fabrication process is the most successful way of producing high quality parts at reasonable cost [3,4,5].

The application of robots in assembly and production of parts from solid materials has been very successful. The non-rigid nature of the composites and the complex assembly processes involved in the fabrication of high precision components have significantly limited the application of robots in composite part manufacturing. Efforts at automation have mainly been directed at meeting the particular requirements of the composite manufacturer. Consequently, system cost has remained high due to the specialised nature of the machines [4,5]. Robotic systems for the assembly of composite parts require a high degree of intelligence, advanced sensory devices and a range of manipulators to meet the challenging requirements of handling non-rigid materials.

The aim of the paper is to focus on research issues which must be addressed to make the sensor-based robotic lay-up cells of the future technically feasible. In particular, detailed consideration is given to sensory handling of the non-rigid materials and the sensory requirements for in-process monitoring and control of the assembly operations. Application of knowledge based robotic systems to handling of fibre composites are briefly outlined. Sensory requirements for detection of manufacturing defects of the consolidated finished product are also considered.

2. CHARACTERISTICS AND APPLICATIONS

A composite material can be defined as the combination or fabrication of several materials acting together to provide properties unobtainable by the individual constituents acting alone. Their combined action is frequently of a different order of magnitude, or of quite a different character, from the behaviour of the components selected from the basic material or from the different materials [1,6]. Commonly used long-fibre advanced composites are glass-fibre, carbon-fibre, and aramid-fibre reinforced primarily with epoxy resin or sometimes with phenolic or polyester resin matrices [1,7]. They have excellent mechanical and thermal properties combined with a great potential for weight reduction and for increasing system performance. Typical advantages of advanced composites include:

. Reduced weight
. High resistance to corrosion and erosion
. Improved fatigue resistance
. High strength-to-weight ratio
. Low thermal expansion
. High resistance to environmental factors
. Good silencing properties
. High electrical resistance
. Permeability for X-rays and microwaves
. Reduction of assembly costs using integral structures
. Increased system performance.

The main reasons for use of composites are weight reduction of the component, potential of designing the component from first principles, thus optimising strength and stiffness behaviour. Higher material costs of advanced composites can be compensated for by integral construction which considerably reduce the assembly tasks required for structures with, for example, a large number of joints. Recent reports on Airbus A310 fin box project have shown that by substituting conventional metal with composites the quantity of detailed parts is reduced by a factor of 22 and the number of fasteners by a factor of 8 [5,7]. In addition to saving considerable assembly cost a reduction in weight of approximately 23% has also been achieved. Figure 1 shows a cost comparison between conventional and composite fin box manufacturing. It should be noted that the actual cost of robotic part production increases while the overall cost of automated production and assembly is reduced from 85% to 65% of the total cost.

Another reason for the adoption of advanced composites is the enhanced system performance. Composites are now beginning to play a significant role in the design of new generation of military and civil aircraft. The design of the high performance experimental X-29 aircraft incorporates forward-swept, computer controlled wings which must be light-weight and also able to resist high-torsion. This is only possible with the application of advanced composites.

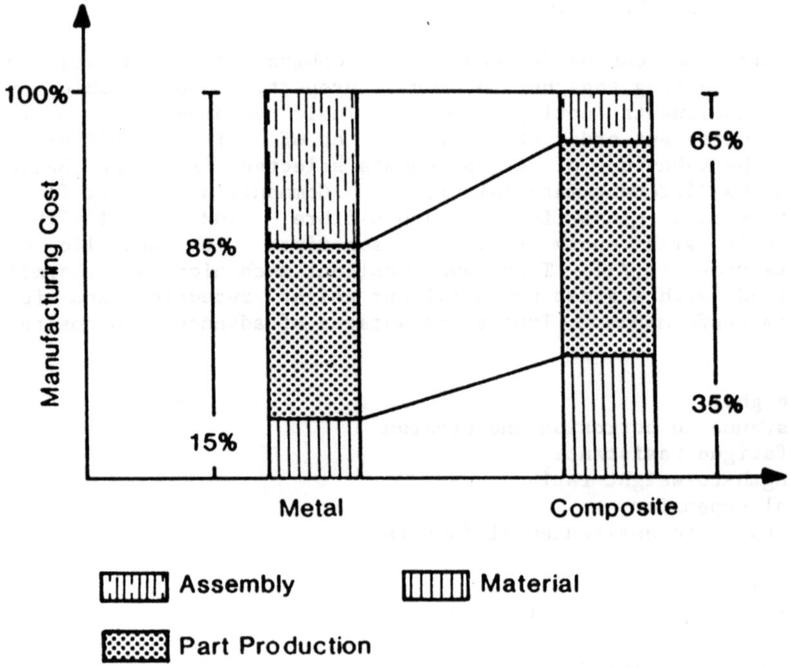

Figure. 1 Cost comparison - metal/composite fin box - After Sarh B. (1984)

A new mode of propulsion, known as propfan or unducted fan, has been under development since 1975 [8]. It combines the efficiency of propellers with ultra-high air by-pass ratios to achieve increased thrust, higher speeds and altitudes than is possible with conventional propellers. A plane can save more fuel in climb and decent than at cruising speeds. Thus, propfan is primarily aimed at the mid-range aircraft, where it is estimated to be up to 25% more fuel efficient. As can be seen in Figure 2, propfan blades are shorter, thinner, wider and more numerous than conventional propeller blades.

Fig 2. Unducted Fan Engine

Ideally the blade material should be light, free of fatigue limitation, with high strength to weight ratio and good damage tolerance. The adoption of a long fibre composite construction comes close to satisfying most of these requirements. With composite material it is also possible to vary the lay-up of laminates within the blades to arrange the natural blade vibration frequency so that no resonances occur in the normal range of the engine running frequencies. It is very difficult to achieve this blade characteristic with conventional metal designs [8].

3. TYPICAL HANDLING TASKS

Two types of basic materials are available to component manufacturers. Fibre composites are normally supplied to manufacturers in rolls of fibre pre-impregnated with resin and partially cured (called prepreg tape). Prepreg tapes are non-rigid, the fibres being either unidirectional or woven. They are available in a limited range of width and length sizes. The tape is covered by a layer of protective plastic or paper film to prevent the roll from sticking to itself and also to avoid contamination in manual handling. For maximum longevity prepreg rolls often require storage at low temperatures. This type of material usually allows closer control of the resin during manufacture than the dry lay-up process where the resin impregnation is carried out, after the component lay-up, by the part manufacturer. However, if the prepreg rolls are not available in longer lengths and broader widths to suit the application, then the number of gaps in the final product may increase the void content [3]. In the dry lay-up process the raw materials are rolls of dry fibre, again either unidirectional or woven. Often, the material supplier applies a special thermoplastic binding net to one side of the tape and covers it with a layer of protective backing film. The three layers are lightly bonded together by a heating process before the tape is rolled up. This process is sometimes carried out by the component manufacturers rather than the material suppliers. The backing film gives the fibre tape a certain degree of shape stability during component lay-up and protects the fibres from fraying at the edges. Dry fibre rolls are cheaper than prepreg tapes. Many component manufacturers have now acquired the technology to accurately control the resin injection process and to produce high-quality parts more economically.

Manufacturing solid composite components from non-rigid prepreg or dry rolls involves two distinct handling processes. As shown in Figure 3, one group concerns handling of non-rigid materials, such as cutting, pick-up, lay-up (either in a mould or flat surface), forming and debulking. These operations occur before the injection of resin in the dry lay-up process and curing in the case of prepreg. The second group, represents solid cured parts and includes trimming, drilling, and finishing.

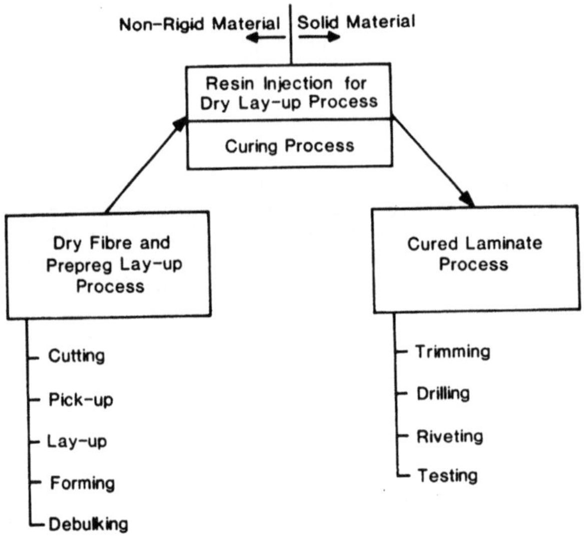

Figure. 3 Manufacturing process involved in composite part production - After Sarh B. (1985) modified

Handling tasks involved in the processing of solid composite parts are not significantly different from those used for metals. Therefore these will not be considered here in detail. However, the tasks involved in handling the non-rigid material present challenging manufacturing problems particularly in fabrication of precision components, often required by the aerospace industry [4]. Some composite handling difficulties are common to the processing of most non-rigid products in garment and shoe industries.

These are cutting, pick-up, destacking, and some lay-up and manipulation tasks. The tacky nature of composite material, the accuracy and the number of laminates (up to 100 or more) required to fabricate a flawless solid component, mean that the difficulties of handling composite materials are greatly magnified. For example, removal of the backing film from the prepreg or fibre layer without fraying the material is a difficult and time consuming manual handling task. Furthermore, the assembly operation is usually carried out in a clean room under controlled atmosphere with protective clothing to remove dust and contamination. Gloves must be worn by the assembly staff which tends to impede their manipulative action to a certain extent. The following example will serve to demonstrate some of the handling operations involved in the dry lay-up and manufacture of an advanced composite propeller blade [8].

In manufacturing terms the process involves the following distinct operations:-

(I) Ply Cutting - The basic fibre materials are unidirectional carbon, to form the main load carrying members, and $\pm 45^{o}$ glass, to produce the aerodynamic skin. The required shapes are usually designed using CAD systems and cut from rolls of dry carbon and/or glass fibres or woven cloth to form plies. These are then delivered to the next operation. Handling operations involved normally include:

- (a) cutting cardboard templates of the ply shapes.
- (b) positioning the fibre or prepreg composite on the cutting table.
- (c) using the cardboard templates and reciprocating knife to cut the required plies.
- (d) taking the plies off the cutting table to the next operation.
- (e) removing the waste from the cutting table.

(II) Preform Assembly - Involves accurate lay-up of a number of plies (up to 12) of various sizes, shapes and fibre orientations to assemble a range of flat geometries (called preforms) shown in Figure 4.

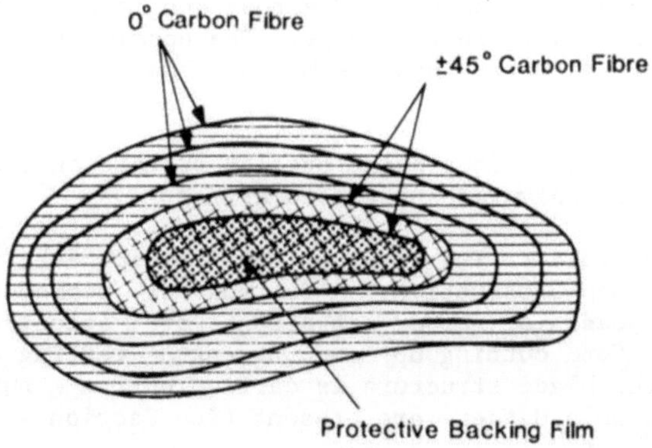

Figure. 4 Filler preform assembly

In blade assembly usually three main preforms are used, namely skin, wrap and filler. The dry assembled preforms can be stacked and stored for unlimited periods at room temperature, for later use. In detail, the handling tasks include:

(a) positioning of the ply on the table.
(b) removal of the protective backing film.
(c) positioning of the next layer on the exposed surface of the first ply. Accurate positioning usually involves cardboard template matching.
(d) application of heat and pressure to activate the inter-laminate adhesive force and achieve bonding with disturbance free fibre surface.
(e) repeat steps (b) to (d) until the required preform layers are assembled.

At this stage the preform assemblies are still non-rigid and the backing film on the top layer is not removed.

(III) Blade Assembly - The preforms are assembled into the blade mould in the correct sequence. The lay-up arrangement is usually ordered in skin, wrap, filler sequence. Handling tasks include:

(a) lay-up skin preform into the half-mould.
(b) remove protective backing film.
(c) use cardboard template matching to accurately position preforms inside the mould.
(d) build up the required number of skin preforms. Apply pressure and heat to remove disturbance and achieve partial preform bonding after each lay-up process.
(e) repeat steps (a) to (d) for wrap and filler preforms.
(f) integrate blade root assembly into the lay-up.
(g) complete skin preform lay-up.
(h) cover the lay-up by the other half of the mould and fasten the two mould halves.

(IV) Resin Injection and Curing - The mould is mounted in a vertical position with the blade root at the top. To reduce resin viscosity during injection, the mould is heated. The root end is connected to a vacuum supply and the tip to a resin feed pipe. The epoxide resin is then fed at a precisely controlled rate into the mould. The blade is then cured and deflashed.

(V) Erosion Protection - To provide erosion protection the blade is spray coated all over with polyurethane elastomer.

(VI) Quality Control - Two methods of testing are employed, namely destructive and non-destructive. To monitor reproducibility of the manufacturing process a number of blades are selected on a statistical sampling basis, for cutting up and strength testing. Non-destructive evaluation of the blade structure is carried out in a number of ways to ensure no unacceptable defects are present (see Section 4 (d)).

Natural frequencies of each blade are evaluated using electronic techniques to measure the fundamental flexural and torsion frequencies.

4. STATE OF THE ART IN AUTOMATED FABRICATION

Figure 5 shows a typical clean room for the manufacture of advanced composite components for the aerospace, and automotive industries [9]. Current practice of composite part fabrication relies predominantly on skilled manual labour for the manufacturing process, even though automated systems are available to perform the less demanding tasks. This is particularly true in the civil aircraft industry where small production runs and limited range of components are normally required. However, as more aircraft parts are made from composites, the demand for automated means of production will increase. In the USA many companies have been working on the automation of handling and manufacture of composite components for the U.S.Air Force [10,11]. The Grumman Aerospace Corporation, Northrop Corporation, General Dynamics and Lockheed Georgia, are amongst the most active in this field. Recent research publications show that Europe is also pursuing active research and development work in automated composite fabrication [5,7,12,13].

Reduced version of Figure 5 to go in here.

Fig 5. A Clean Room at Darchem Specialised Mouldings for the fabrication of advanced composite components - courtesy of Smye Holland Associates.

The following is a brief survey of the State of the Art from the design stage to testing.

(a) Design and Analysis - CAD/CAM facilities are now widely used to design composite lay-ups and hence components. However, it often requires a great deal of engineering design expertise to create and store a high-performance 3-D component from 2-D composite plies. Systematic methods of representing material geometry are beginning to emerge to facilitate rapid design procedures. Both general purpose as well as the more accurate specialised Finite Element Analysis (FEA) programme are now available to evaluate the component characteristics [14].

(b) Ply Nesting and Cutting - Minimum wastage of expensive composite material requires careful nesting of large numbers of composite ply shapes. Generation of nesting with dense geometric definitions either manually, using miniature patterns, or employing fixed nesting methods, which manually position ply shapes on a CAD/CAM graphic system, is a very time consuming operation. A single nest containing a large number of plies can take several days to complete.

The trend is towards automated nesting systems capable of nesting complex 2-D ply shapes that link up directly with the cutter [15]. Bell Helicopter have implemented a just-in-time automated nesting system for advanced aerospace structures which integrates CAD/CAM graphics, Material Requirement Planning, Computer Aided Planning and Shop Floor Control Systems [16].

Waterjet cutting has been accepted as a method of cutting different composite materials for many applications. Cutting nozzles are small, a few thousands of an inch in diameter, with water pressure in excess of 40,000 psi. When abrasive is added into the waterjet stream, it has the advantage of cutting metallic composites such as boron/aluminium. Die-cutting is another technology which is widely used by composite manufacturers. Laser technology makes an attractive non-contacting cutting system which has the advantage of cutting flexibility when encountering curves or complex ply shapes [17]. Westinghouse Electric Corporation have recently produced a flexible robotic prepreg composite cutting system [18]. The system employs the effective integration of just-in-time manufacturing, automatic nesting, automatic scheduling, robotic cutting with either ultrasonic or reciprocating knife, automatic material loading.

(c) Automated Lay-Up - Sensor-based handling devices are the most significant elements of flexible robotic assembly systems for the lay-up and manipulation of non-rigid composite plies. In recent years many systems have been developed to tackle the handling problems of composite part assembly [3,5,8,10,12]. Basically, these systems are only capable of flat or moderately profiled lay-up assemblies.

Research workers at the Queen's University of Belfast in collaboration with Short Bros, have developed an experimental robotic lay-up cell to investigate the flexible automation of prepreg lay-up operations. They have developed a vision-based adaptive light-weight gripper containing a number of cells within which air pressure or vacuum can be individually controlled to match each particular ply. In pick-up mode the vacuum is used to lift the ply and the surrounding positive pressure to ensure that adjacent plies are not disturbed. In the lay-up mode the active vacuum cells are rapidly turned to positive pressure to overcome the adhesion of the prepreg plies to the gripper surface at room temperature. Preliminary experiments with stacking of flat multi-layer plies have been very encouraging. Currently, the gripper surface is flat but it would be possible to fit profiled grids to allow lay-up directly into a component mould. Furthermore, rapid change over from cell vacuum to positive pressure levels facilitates modulated control of the cells. This, together with controlled release of the ply, from one end to the other, directly into the component mould would ensure that pockets of air are not trapped between successive plies.

In the Airbus Fin Assembly extensive use is made of robotic vacuum handling and bandaging devices. Adaptive vacuum is used in the pick-up and lay-up operations. A spring loaded vacuum drum is used to bandage prepreg tapes around aluminium core modules. The drum is spring loaded and relies on greater adhesive force between the prepreg and the aluminium core than that beween the prepreg and the drum to apply the prepreg bandage around the module core. Six robotic tape-laying heads each capable of accommodating a different width tape role have also been developed for this application.

Designers at Georgia Institute of Technology have proposed two automated lay-up processes for discrete and continuous flat sheets of up to 8 layers of PEEK/graphite fibre composite. Both processes are based on currently available technologies and are aimed at aerospace and automative industries. The lay-up processes effectively eliminate manual handling of the rolls following delivery to the lay-up system.

State of the art in handling of non-rigid fibre composite plies tends to favour the use of vacuum, air pressure and mechanical grippers. However, a recent report has suggested the use of electroadhesion for carbon fibre ply pick-up application [19]. Preliminary investigations with an electrostatic gripper designed, mainly for application in textile industry has proved successful in lifting carbon fibre plies [20].

The handling mechanisms described above rely on manual separation of the backing film from the ply. Researchers at the University of Durham have developed a rolling type manipulator for handling non-rigid materials such as textiles and fibre composite plies [21]. The manipulator is fully instrumented and employs an adjustable length pin arrangement to engage the material. A set of strain gauge bridges resolve the roller reaction forces pertinent to ply engagement and separation, by means of continuous peeling and wrapping of the ply around the roller head. The manipulator has proved to be very effective in removing the backing film from prepreg and dry fibre plies. When the adhesive force between the ply and the underlying surface is greater than the peeling force of the manipulator, the ply is not unduly disturbed. (see Figure 6). Otherwise, a sliding S shaped clamp is mounted close to the manipulator to protect the ply against local disturbance during the film removal operation. (see Figure 7).

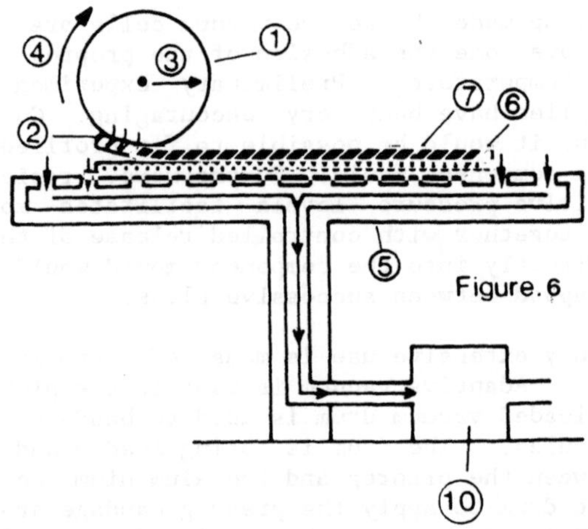

Figure. 6 Automated removal of the backing film on vacuum table

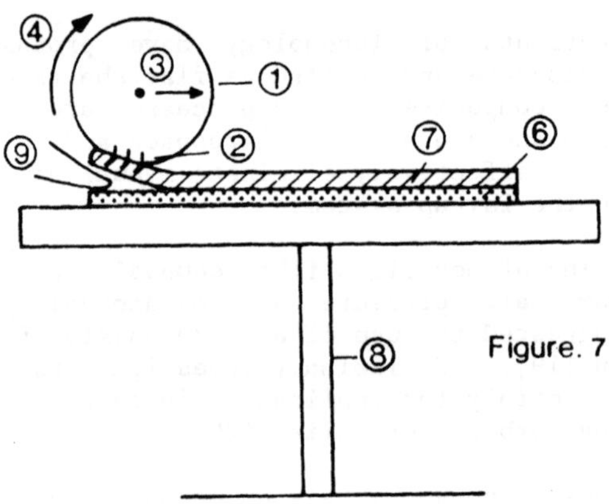

Figure. 7 Automated removal of the backing film on ordinary table

① Rolling type fully instrumented robotic manipulator

② Adjustable oblique pin arrangement

③ Translational movement

④ Rotational movement

⑤ Vacuum table

⑥ Dry fibre or prepreg ply

⑦ Protective backing film

⑧ Table without vacuum facility

⑨ S-shaped sliding clamp

⑩ Vacuum Pump

Current research is aimed at developing a fully automated film removing manipulator.

Clearly, preliminary investigations are now being carried out to provide solutions for automated lay-up of composite components. However, it will be sometime before cost-effective robotic sensory handling technology is available for the fully-flexible automation of composite part production processes. Semi-automated assembly where human operators still play a key role in the fabrication process is perhaps more cost effective and of immediate interest to manufacturers of composites. An interesting approach to human centred automation of the lay-up process is that being developed by Assembly Guidance Systems [22]. The Computer system projects reference laser graphics onto the lay-up mould which act as templates of light to speed up the manual lay-up operations. It also projects lay-up instructions such as the type of ply required, fibre orientation etc. at each step in the assembly process. The lay-up reference graphics are corrected for distortions resulting from projections on complex curved moulds. The system replaces the tedious and time consuming practice of template matching, string or registration pins used in the assembly operation while reducing lay-up errors and the amount of paperwork associated with it. The laser projector may be fitted with a video camera to provide a feedback to either the computer or the quality assurance inspector on the accuracy of the lay-up.

(d) Sensors for Flaw Detection - The high number of processes involved in the fabrication of composite parts makes many manufacturing defects possible. These include delamination, void, porosity, inclusion of moisture and other foreign particles, incorrect fibre alignment or broken fibre, variation in resin content and improper resin cure [23,24]. The extent to which defects can be tolerated will depend upon the effect they have on the performance of the component. This will vary widely, depending on the part geometry and structure, location of the flaw and the type of application for which the part is designed. Therefore, it would not always be a good practice to classify defects into acceptable or unacceptable, critical or non-critical, for the general case.

Several Non-Destructive Testing (NDT) Systems based on a range of sensors are available for identification and characterisation of defects. Increased use of composites has led to a search for more accurate and faster techniques for detecting manufacturing flaws.

Current techniques of composite NDT include: ultrasonic, X-ray radiography, computer tomography, acoustic emission and acoustic microscopy, holography, thermography. A few of these techniques are widely used by composite manufacturers. Ultrasonic testing is perhaps the most popular technique, since it detects most commonly encountered flaws. Radiography is often used to identify defects such as fibre mis-alignment and foreign inclusions which ultrasonics cannot detect.

5. FUTURE AUTOMATION NEEDS

There is a concensus of opinion that before cost-effective and flexible composite automation systems are applied in the real-world, major research efforts need to be directed at developing enabling technologies for a range of hardware, software and integration problems. Some of these problems are specific to handling and fabrication of composite parts. Others are of general interest to robotics and advanced flexible manufacturing cells. In the case of the robotic lay-up of composite parts these include:

(a) Component design for automated manufacture
(b) Automation of handling
(c) Sensory development
(d) Intelligent robots

(a) Component Design

A recent study has concluded that using expensive robotic techniques to imitate manual lay-up methods will severely limit the advantages that might be gained by automation [25]. Components should be designed for the ease of lay-up and automated manufacturability. The ultimate objective of the researchers and engineers should be to achieve flexible automation systems capable of operating with minimum human intervention. This will not be an easy task, especially in the aircraft industry where each new component needs to be approved and licenced by the appropriate authorities.

Fabrication of the Airbus horizontal stabilizer has successfully addressed the critical design issues. The design employs repetitive modules which are then integrated into a complex structure. The project has shown that it is possible to establish an economically feasible system capable of producing parts at a faster rate than manual fabrication.

In the short-to-medium term it may not be possible or desirable to totally eliminate the role of human operators from the assembly process. In that case, the design should aim at producing a series of composite preforms of relatively simple profiles with which to make up the final integrated structure. Steps involved in assembling the preforms require a large amount of labour. However, in the near future, it should be possible to build on the existing automation technologies to establish flexible manufacturing systems capable of fully automated preform assembly. This approach will enable a gradual reduction in the amount of manual labour from the complete component fabrication process.

Research work in this area needs to be aimed at developing a general mathematical modelling technique for transforming 3D solid component structures to 2D ply shapes and vice versa. Such a transformation technique would speed up the design procedure for maximising the number of preform assemblies in each component.

(b) Automation of Handling

Complex handling tasks involved in the lay-up of non-rigid dry fibre or prepreg plies into complex component moulds will probably require a range of robotic ply transfer mechanisms, manipulators, film removing devices, profile interfacing plates, integrated pressure rollers and heating devices, cutters, etc. Some of the handling tasks such as the protective backing film and paper removal require delicate manipulations to avoid damaging the ply, while forces in excess of 200 N have been applied in other composite lay-up operations [5].

The composite lay-up cell of the future will probably include more than one robot sharing a tool crib containing a number of end effectors and mechanical fixtures. The range and nature of handling devices will ultimately be dictated by the complexity of composite component structure and the type of fibre or prepreg material used in its lay-up. The need to iron out wavey disturbances in the lay-ups and eliminate inter-laminate voids in a complex surface profile tends to favour rolling devices. Furthermore, the dexterity of such manipulators may be increased with improved designs, for example, by incorporating a degree of mechanical programmability.

(c) Sensor Development

Flexibility of a robotic lay-up system is a function of its manipulator capabilities and sensory feedbacks as well as the efficient integration of its hardware and software. Devices for the handling and manipulation of non-rigid plies require a combination of the following sensors to obtain local processing information.

(i) Force, torque, pressure and position sensors to control contacting, grasping and pulling forces at each particular position in the lay-up process.

(ii) Tactile array sensors to measure the distribution of forces over the gripper and the lay-up surfaces.

(iii) Ultrasonic or acoustic imaging sensors to derive proximity information on the profile of a complex component structure.

General purpose inexpensive sensors such as those in (ii) and (iii) above, that are capable of tackling the complexities of the manufacturing environment, are currently not available and would require further development.

Vision systems form an essential part of any flexible manufacturing system, particularly in the lay-up of non-rigid plies where many processing variables have to be precisely monitored and controlled. Flexible robotic lay-up systems require high-resolutions, grey level vision sensing systems, equipped with multi-camera inputs to perform global monitoring and control of the lay-up process. Such a sensing system would have the ability to detect in-process changes in the ply such as surface disturbances, accidental folding, improper overlapping as well as monitor the structural details of the assembly process. Ideally, it would also interact with the robots and other sensory devices to facilitate the unmanned composite fabrication cell.

However, processing, information extraction and interpretation of data from complex sensory systems is still in its infancy [26]. Existing grey-level vision systems are too slow and expensive to be widely used in industrial applications. Considerable research and development in hardware and software is required to make three-dimensional vision systems commercially available in the next decade or so.

Binary vision sensing offers limited capability at a reasonable cost without extensive efforts towards structured lighting. It would be possible to equip the existing composite ply cutting systems with an ink marker to clearly idendify the contour of the next layer in the preform stack. Binary vision systems integrated into robotic workstations would then be able to lay-up flat composite preforms with existing technology in an environment without structured lighting.

(d) Intelligent Robots

Fabrication of precision solid components with elaborate curvatures from non-rigid composite materials involves many complex and unpredictable manufacturing problems for which existing robotic technology is not well suited. Intelligent, knowledge based robotic control systems is an active research field which is aimed at better device integration and using sensory information to achieve a much higher level of intelligent system behaviour. Such systems are ideally suited for processing non-rigid material since they would be able to cope with uncertain situations, learn from experience and recover from errors. An excellent overview of the current knowledge based robotics is given by Williams and Rogers [27] and practical automation work in intelligent robotics in processing of non-rigid material in Bristol is described by Khdodabandehloo [28,29]. Extensive research and development work in sensory based intelligent robots at Carnegie-Mellon University is reported by Sanderson and Perry [30].

6. CONCLUSIONS

Sensory handling of fibre composite plies has been identified as the fundamental element in the automated lay-up of composite components. A review of the state of the art has shown that, provided parts are designed for automated manufacturing, many of the assembly operations in the lay-up of composites can be automated with further development of existing technology. However, considerable research and development efforts are needed to establish fully automated flexible robotic lay-up cells capable of unmanned assembly operations.

7. ACKNOWLEDGEMENTS

The author would like to acknowledge the financial support of ACME Directorate of the Science and Engineering Research Council in research projects into the automation of handling and production of components made from non-rigid materials at Durham. Thanks are also due to Dr.T.V.Parry for his assistance in the preparation of this paper.

8. REFERENCES

1. King, R.L., "A Production Engineers View of Advanced Composite Materials", Materials and Design, Vol.3, August 1982, pp 515-522.

2. "News on SRI International", Advanced Composite Engineering, Vol.1, No.12, Winter 1986, pp 2.

3. Colton, J.S., et al, "The Automation of Lay-up and Consolidation of PEEK/Graphite Fibre Composites", 32nd Int. SAMPE Symposium, April 6-9 1987, pp 1138-1151.

4. Stone, K.L., "Automation in Composites Processing", 29th National SAMPE Symposium, April 3-5 1984, pp 1489-1498.

5. Sarh, B., "Robotic Applications to Automated Composite Aircraft Component Manufacturing", SME Technical Paper, MF 85-506, 1985.

6. Amini, F. and Atack, P.A., "A Survey of Composite Materials", Sheet Metal Industry, Jan. 1974, pp 7-15.

7. Maskow, J., "New Production Techniques for CFRP Components in Airbus Production", SME Technical Paper, EM87-137, 1987.

8. McCarthy, R., "Manufacture of Composite Propeller Blades for Commuter Aircraft", SAE Technical Paper, 850875, 1985.

9. Twist, B., "Out of the Bag" Advanced Composites Engineering, Vol.3, No.1, Spring 1988, pp 19-21.

10. Allison, H.B., Pitman, W.A., "Future Trends in Technology and Automation", AIAA International Meeting, Paper No.80-0922, May 1980.

11. Dreger, D.R., "The Challenge of Manufacturing Composites", Machine Design, October 22, 1987, pp 92-98.

12. Wright, E.J., et al, "Robotics Handling of Composite Materials in a Lay-up Cell", U.K. Research in Advanced Manufacture, Proc.of I.Mech.E., London, Dec.1986, pp 39-46.

13. McGeehin, P., "Composites in Transportation - Design and Current Developments", Materials in Engineering, Vol.3, April 1982, pp 378-387.

14. Rouse, N.E., "Optimising Composite Design", Machine Design, Feb.25, 1988, pp 62-66.

15. Singer, S., "Composite Die Cutting Systems", SME Technical Paper, EM86-713, 1986.

16. Bates, J., "Just-in-Time Composite Ply Nesting", SME Technical Paper, MS87-131,1987.

17. Mello, M.D., "Laser Cutting of Non-Metallic Composites", SME Technical Paper, EM87-138, 1987.

18. Guha, J.K., "Flexible Robotic Prepreg Composite Cutting System", SME Technical Paper, MS87-138,1987.

19. Monkman, G.J., and Taylor, P.M., "Electrostatic Grippers, Principles and Practice", Proc.Int.Symposium on Industrial Robots, April 1988, ISBN 0-948507-97-7, pp 193-200.

20. Taylor, P.M., - Private Communication, May 1988.

21. Sarhadi, M., "A Novel Manipulator for Automated Handling of Flexible Materials", 2. Duisburger Kolloquium Automation und Robotik, July 1987.

22. Blake, S., "Laser Guidance for Hand Laid Composites", SME Technical Paper, IQ 87-134, 1987.

23. Korane, R.J., "Spotting Flaws in Advanced Composites", Machine Design, Dec.10, 1987, pp 99-104.

24. Parmar, M., and Boyle, G., "Finding the Hidden Flaw", Advanced Composite Engineering, Vol.3, No.1, March 1988, pp 14-16.

25. Krolewski, S. and Gutowski, T., "Effects of the Automation of Advanced Composite Fabrication Process on Part Cost", 18th Int.SAMPE Technical Conference, Oct.7-9, 1986.

26. Wright, P.K., and Englert, P.J., "Sharpening the Senses of Industrial Robots", Mechanical Engineering, May 1986, pp 58-63.

27. Williams, D.J. and Rogers, P., "Opportunities for Knowledge-based Systems in the Control of Manufacturing Automation", 2nd UMIST/ACME Workshop on Advanced Research in Computer Aided Manufacturing, UMIST, UK, 7-8 Jan.1986, pp 1-14.

28. Khodabandehloo, K., et al, "Intelligent Vision Guided Robots for the Food Industry", The International Symposium and Exposition on Robots, Sydney, Australia, 6-10, November 1987.

29. Khodabandehloo, K., et al, "Robots with Artificial Vision and Intelligence", Advanced Robotics Programme (ARASA) International Conference on Nuclear Robotic Technologies and Applications - Present and Future, 29 June - 1st May, 1987.

30. Sanderson, A.C., and Perry, G., "Sensor-based Robotic Assembly Systems: Research and Application in Electronic Manufacturing", Proc.IEEE, Vol.71, No.7, July 1983, pp 856-871.

CHAPTER II

GARMENT MANUFACTURING

A CHALLENGE TO DEVELOP FULLY AUTOMATED GARMENT MANUFACTURING

By Henry A. Seesselberg
Director of Advanced Technology Operations
Fashion Institute of Technology
New York City, USA

In 1987, the Apparel Research Committee of the American Apparel Manufacturers Association produced a report[1] on the state of the art of limp fabric handling devices. It would be easy for me to, at this time, review the whole report. But many of you may have seen this report or can obtain a copy from the AAMA and, as my message at this conference is to principally present a blueprint for the future rather than dwell on of where we've been, I would like to just review the reports conclusions and then I will speak to that future by presenting a challenge to the apparel industry.

These conclusions are stated in the referenced report quite forcibly as follows: "There have been very few organized projects which focus on the pick-up of a single ply from a cut stack. For the most part, work has been done either by individual inventors or by a single machinery manufacturing company in an effort to devise, develop, and market it."

"The automotive industry takes pride in the progress they have made in automation. T.V. advertisements show a dozen or so robots performing repetitive operations such as welding with sparks flying. This is very impressive to show the progress made in that industry. However, this success has not been achieved in the apparel industry basically because of the characteristic of limp fabric and the difficulty in handling this material. Although some pick-up devices have had limited success in specialized areas of apparel, they are not universally utilized in all segments of the industry."

NATO ASI Series, Vol. F 64
Sensory Robotics for the Handling of Limp Materials
Edited by P.M. Taylor

"Considering the world-wide mechanical/electrical progress and development, one would think that limp fabric handling techniques and devices would have made much more progress over the years than the current state of the art. There seems to be some negative influence, economic or otherwise, which has greatly limited R&D efforts in this area. However, the fact remains that one of the basic reasons for the stunted growth of automation in the apparel industry is the lack of progress in limp fabric handling, beginning with the ability to pick up a single ply from a stack of cut parts."

"Considering the recent rush of international trade and competition for the cheapest and lowest price, logically, more effort would have been directed toward progress in limp fabric handling as a means of further reducing labor content. Work aids can no longer carry the cost reduction required to remain competitive."

The report then briefly discusses the problem of cheap labor competition and states, "We believe that a vital requirement to overcome cheap labor is by extensive use of automation and full automation is only possible if reliable robotic material manipulators are widely available."

My challenge to you, then, is to find a way in which garments or parts of garments may be mechanically manipulated, without the use of human hands, sight or intelligence, to allow automatic assembly and finishing of a garment. I guess I should really say "re-challenge" because numerous attempts have been made to do just this and with only limited success. But what is new in my challenge is to approach the problem in a different way, starting at a different place than has been done in the past so that we can, in the not so distant future, transform the apparel industry from a predominately labor intensive one to one which mass produces products by extensive use of totally automatic equipment.

For a moment, however, let us look at where we have been. Traditionally, the development of devices and techniques to automatically feed fabric parts to seaming machines, manipulate them around the needle, and remove them has started at the sewing machine. One watches what the operator does with her hands and tries to mechanically duplicate the action and result. This "inventors" approach, at the end of the process, has not produced many breakthroughs in the fabric handling art except for very narrow applications. I submit that this is because the inventors approach (really in many instances, a tinkerer's approach) has been wrong. If other industries used this approach we would still be buying cars with bolted frames and radios with wired valve chassis.

What we must do is emulate other industries' approach to the automated assembly problem. This means starting at the beginning of the manufacturing process, i.e., the product design, and use a scientific engineering approach instead of that of a gadgeteer.

In any product design, a material is chosen or evolves which fits the requirements of the products end use. Auto bodies use sheet steel and molded plastic; furniture uses wood, fabric and plastic; books are made of paper, plumbing is iron, copper, or plastic; windows are glass or plastic; and clothing is made of woven or knit natural fibers or plastic. Fastening and joining methods have been developed to also match the final acceptability in the end product. Sheet steel is riveted, welded, or bolted, wood uses nails, staples, and glue, paper is stapled, glued, or sewn, pipes are soldered, threaded, or glued; glass is fastened with points or glue and clothing is sewn with thread and assembled with adhesives. And, they all serve the product design and manufacturing techniques to obtain that product design. Our particular interest, clothing, has materials that flexibly mold to the body shape and movement, feel soft against the skin, are washable, and improve the look of the human body while protecting it from the environment.

But! These material requirements of <u>garment</u> product design are the very reason that the manufacturing process is still largely manual. The complexity and difficulty of handling in assembly seems to defy full automation. Yet, we must find a way or we will see our industry go to others who either can supply the cheap manual labor or perform the engineering necessary for automatic garment assembly.

Let me give you an illustration of where we are in truly engineering a solution to fully automatic garment production. Suppose that we have obtained the services of one of the best mechanical engineers in your country. We will ask him or her to design a machine for us to do the following:

A) Pick up and move two pieces of material through a single needle punching machine in a precise pre-determined path. The material must not move while the needle is penetrating it.

B) Remove the finished assembly and place it in a position to be picked up by the following machine.

Sounds like a real simple requirement to us, doesn't it? In fact, its done in many places for shirt fronts, labels, and other simple assemblies. But we want a machine engineered for the ultimate in reliability, versatility, and economy, one which rivals the performance of such machines in the other major manufacturing industries.

Our bright engineer would no doubt ask the following initial questions:

A) What are the dimensions of the parts and the range of these dimensions from product to product?

B) How fast must it be done?

C) What are the physical characteristics of the material, i.e., hardness, malleability, tensile strength, surface roughness, bendability, porosity, etc.

D) What must the finished assembly look like?

How would we answer him or her.? Probably like this?

A) Dimensions? I don't really know. When flat one part is 13 x 16 inches and the other 12 x 12 inches with a curve on one side. However, when picked up by a corner these dimensions may stay the same, if stiff denim, or assume an entirely different shape with complex three dimensional components (wrinkles to us). But, when picked up in the center, they assume a completely different set of dimensions. Also, they may change dimensions with time and humidity.

B) How often? That's easy -- let's say every ten seconds.

C) Material characteristics? Well one piece has a medium to good hand, a stiff drape and it is slick or slippery. The other is a "hard" fabric cut on the bias. But as far as those other quantitative engineering terms such as roughness, strength and maleability, I really can't say.

D) Finished appearance? It has to look good with no seam "grinning" or "pucker". Also, the joining seam has to have some "easing" and the ends must match "exactly", not 12" + or- 1/16", but "exactly".

Our super engineer's next statement would probably be "It can't be done with such an ill defined material and desired end result. I must know the specifications of the material and end product more quantitatively."

Yet, we do this assembly operation all the time, don't we? 35,000 factories in the U.S.A. alone do it every day. But, we do it with human operators instead of machines because, at present, they are the only mechanisms capable of intelligently, properly and completely manipulating this material. And, it would seem as if, unless we are able to quantify the characteristics of the material with which we are to work so our mechanism engineers can design automated equipment properly, we are doomed to continue on this high labor content path.

This valuable engineer would, no doubt, seek meaningful employment in another industry where the rules of the game were somewhat more precise affording some hope of success. In almost any other manufacturing industry he or she chooses the materials that are described by quantitative values, particularly size and shape. This allows for predictable behavior in moving through the assembly process. Sheet metal can be precisely described as to its length, width, thickness, hardness, marketability, bendability and roughness. Pick it up, drop it, lay it down, turn it, and it retains its shape. Join two pieces together and the resultant dimensions are also predictable. Wood, paper, plaster and glass also have this predictable stability. But fabric. ..ah, that's another story. Most of us know exactly what hand, drape, and stretch are. Or do we? These are merely vague quantitive ways to describe the feel and look of fabric on the human body. I say that, in engineering terms sufficient for proper design of product and production machinery, we don't know what we are talking about.

Now let's revert to a statement I made earlier, i.e., the one about narrow application breakthroughs. It is true that a number of devices exist which, in a rudimentary way, pick up limp fabric. These include the Clupicker,[2] Polytex,[3] Littlewood,[4] and Walton[5] devices, which are some of the better known names in cloth pick up technology.

Figure A depicts the Clupicker, which uses a pinching technique very similar to how human fingers pick up the top ply of cloth. If that's all that had to be done, it works fairly well. But all it can do is pick up one ply, not handle parts or garment assemblies.

Figure B shows a needle pickup device made by Polytex. It is, as are the others are, made with extreme precision and therefore high cost, particularly if several are needed. It also can only pick up one ply of fabric.

Figure C is another needle device by Littlewood which is really nothing more than a multiply variation of the needle principle used in many attempts at reliable ply pickup.

Figure D illustrates one of the many Walton devices of which only one is now being used experimentally that I know of. The one shown uses a combination of air foil, needles, and suction to effect a "oneply only" pickup. But, it never addresses what you do after pickup. I believe of Draper Labs in Cambridge, Mass. and Singers' work for TC^2[6] in utilizing Walton pickers, of another kind, to actually take a garment assembly robotically through sewing do address this area, but I believe the results are limited.

The AAMA study actually found 47 different device developments of which only about three are now in commercial use.

Yet even these three devices are seldom used and only solve one aspect of the manipulation problem. And they use one or more of the following principles with little concern of quantitatively defined material.

Pick Up Method		Operating Characteristics
Suction	-	The material must be relatively non-porous (but how much non-porous?)
Gripping	-	The material must have stable dimensions if the gripper is to be simple
Friction	-	The dimensions must be stable and the fabric must be stiff
Penetration	-	The penetration device device must not damage the fabric and it must cover a wide area to be effective
Adhesion	-	Adhesive tape or medium must be cleaned often and not stick "too hard"

The above techniques and others, have all been used in some special machinery to manipulate fabric parts. But, they are not generally successful or in widespread use. Why? These are three main reasons:

A) They lack true engineering approach

B) They lack or ignore quantitive material data

C) Equipment manufacturers view such devices as add on gadgets and do not devote the research effort required for A & B above to broadly solve the problem.

All this sounds pretty awful, doesn't it? My God, it looks like a hopeless situation! Some of us occasionally may ask, "Why did I ever get involved in an industry which has impossible material. "Is there no hope?" Don't get discouraged! You all didn't get here by being discouraged. There is a way out for those who see it as a <u>challenge</u>, as a tremendous opportunity. View the prior art as a great stepping stone, not as an indication that it can't be done. Indeed, much of the work in this area up to now is very good and provides much of the basis for further technology.

The report on the subject by the Apparel Research Committee in 1987 states, in part:

> "...It is our collective opinion that this work is vital for the survival of the industry, and continued and intensified efforts should be devoted by the industry to the development of these devices and that the AAMA be constantly apprised of progress in this field. It is also recommended that the Apparel Research Committee recommend and urge a major R&D effort in this area be undertaken by TC^2 and/or any other organization involved in the development of apparel manufacturing technology. This effort should make use of fabric properties evaluations systems (such as the Kawabata Evaluation System)[7] in correlating the mechanical properties of the fabrics with their handling properties."

In recent years we have seen great strides in automatic equipment but they almost always require an operator to load and unload piece by piece and usually require human work steering. The operator combines the power of ten fingers, two eyes, and a complex brain in a constant feedback system to properly do this work. The next great stride must be the reduction of the dependence on this costly operator faculty. It must be done before someone else takes the business away from us. That is the exciting challenge! And it's a scientific challenge, an engineering challenge! Not only of equipment engineering, but fabric engineering. Attack the problem at the start. Define your fabric and design your garment as if our super engineer was to apply his or her technical prowess to designing the machines to handle them. Meet the problem on those terms. Back up and look down the process from the material and design end, not up stream from the sewing machine end. If we are to do that we must produce a fabric specification sheet that looks not like this Fig. E, but like this Fig. F.

Then our machinery specifications must look not like this Fig. G, but like this Fig. H. I truly believe it can be done and it's not as insurmountable as I may have implied at the beginning of my talk. In fact, once we can define the problem in engineering terms we can bring the whole engineering world to bear on it and I believe the solutions will be evident and relatively easy.

Of course, as with any cure, we must watch side effects. Taking the operator out of the loop could also cause the loss of some functions not readily recognized as being done by her, such as inspection and correction of prior process errors like inaccurately cut parts and bad seam margins. Also, the versatility and adaptive control of the human operator is taken for granted and must not be lost. There must be viable trade off between labor cost reduction and capital expense increase. And, of course, reliability of the equipment must be paramount. We must be careful, too, that the requirement of engineered characteristics of fabrics does not slow up the response time of a fabric producer to meet fashion demands. These are all part of the challenge.

In summary, I challenge the whole research and engineering community and the fabric, machinery, and garment producers to look at the solution of automatically manipulating limp fabric as a start to finish engineering problem which requires each step of the process to be quantitatively defined in engineering terms. Modern day scientific and engineering knowledge must be applied to our fabrication specifications as it successfully has been in many other industries. The day of the tinkering inventor in our business should be waning and the technical community should be encouraged by our willingness to play with <u>their</u> rules, but in <u>our</u> ballpark. If we can, in a few years hence, point to limp fabric garment assembly technology as a truly scientific endeavor, we should reap not only the primary rewards of maintaining a large industry in the developed nations, and expanding the equipment business, but also attracting new technical and managerial talent into our industry thereby reinforcing its progress.

What I am asking for is this:

- Define your fabric as an engineering material
- Design your equipment to handle this engineering material
- Design your garments and factories to use this engineering material and equipment
- Combine them all in Computer Integrated Machinery

References

1. Apparel Research Committee of the American Apparel Manufacturers Association: Limp Fabric Handling Devices, April 1987.
2. op.cit.
3. ibid.
4. ibid.
5. ibid.
6. F. Fortess: TC^2 Technology - Squaring off with the competition; Bobbin Magazine, May 1988.
7. S. Kawabata, R. Postle, M. Nina: Objective Specification of Fabric Quality, Mechanical Properties and Performance; The Textile Machinery Society of Japan, 1982.

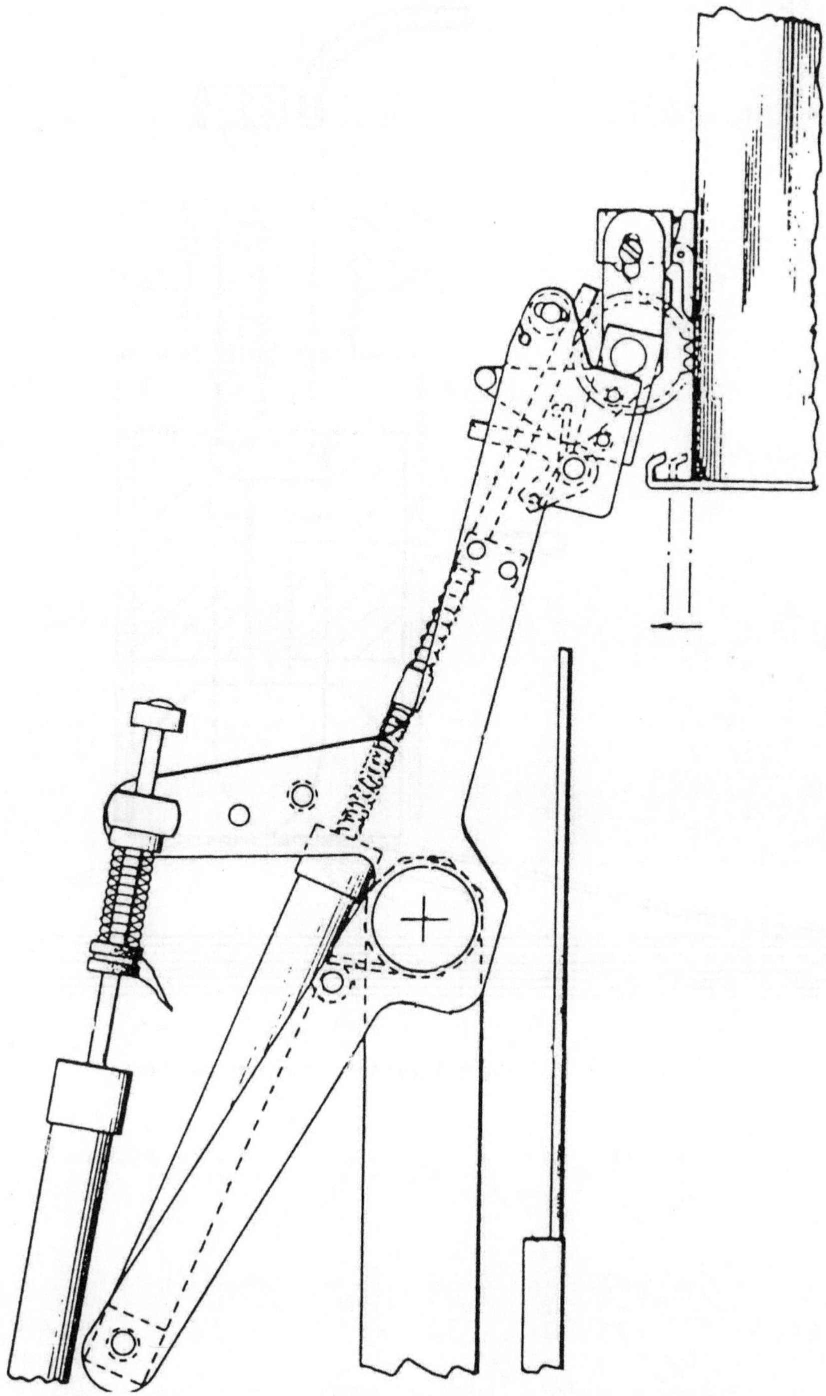

Figure A - Clupicker

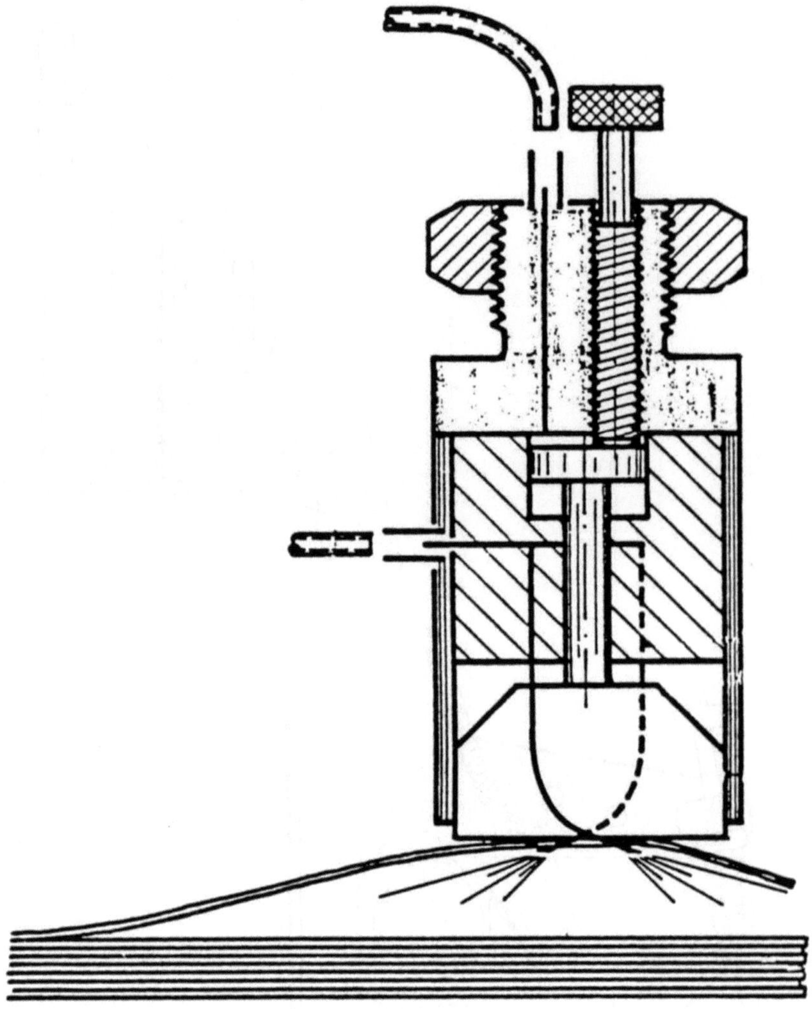

Figure B - Polytex Fabric Pickup Device

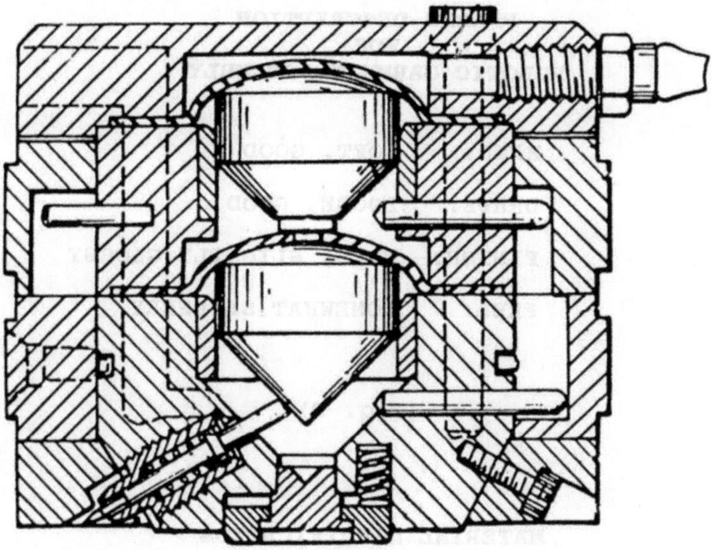

Figure C - Littlewood Fabric Pickup Device

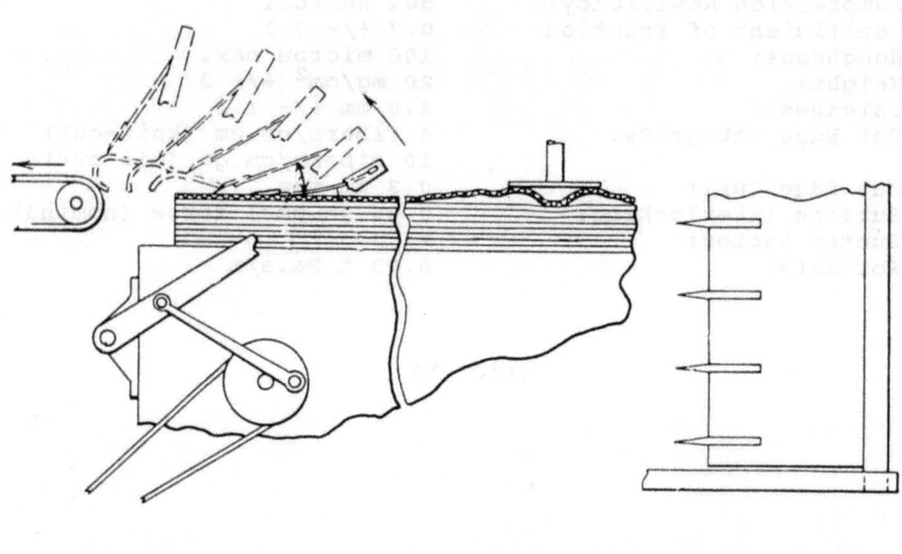

Figure D - Walton Pickup Device

FABRIC DESCRIPTION
FOR
ROBOTIC GARMENT ASSEMBLY

HAND:	SOFT, GOOD
DRAPE:	LOOSE, GOOD
FINISH:	HARD, SLIGHTLY GLOSSY
FEEL:	SOMEWHAT SLIPPERY

(Fig. E)

MATERIAL SPECIFICATION
FOR
ROBOTIC GARMENT ASSEMBLY

Tensile Resiliency:	75% +/- 5
Sheer Rigidity:	0.7 max.
Bending Rigidity:	0.1 max.
Compression Energy:	4 gf.cm/cm^2 nominal
Compression Resiliency:	50% nominal
Coefficient of Friction:	0.7 +/- 0.1
Roughness:	100 micron max.
Weight:	20 mg/cm^2 +/- 3
Thickness:	6.0 mm +/- 1.2
Cut Edge Integrity:	4 fibers/cm.gm (knifecut) 10 fibers/cm.gm (die cut)
Cut Edge Curl:	0.3 cm max.
Surface Interlocking:	0.06 gm peel force (nominal)
Pucker Factor:	2
Porosity:	0.03 K Pa.s/m

(Fig. F)

MACHINE DESCRIPTION
FOR
ROBOTIC SLEEVE MAKING

Allowable Fabrics:	Many suit fabrics (except very slippery, stiff, limp, or porous)
Cycle Time: actual	20 Seconds per piece (approximate - speed depends on fabric type)
Seam Quality: stitch	Pucker minimized (depends on fabric) line straight (if fabric not stiff
Pick Up Reliability:	Depending on fabric and humidity can approach 98%
Edge Placement Accuracy:	Within 1/8 inch for some 60/40 blends (other fabrics may perform differently)

(Fig. G)

EQUIPMENT SPECIFICATION
FOR
ROBOTIC SLEEVE ASSEMBLY SYSTEM

INPUT FABRIC REQUIREMENTS:	Tensile Resiliency	80%
	Sheer Rigidity	0.8
(all maximum except as noted)	Bending Rigidity	0.1
	Compression Energy	5 g ficm/cm^2
	Compression Resiliency	75%
	Coefficient of friction	0.5
	Roughness	85 micron
	Weight	15 mg/cm^2 minimum
	Thickness	8 mm
	Cut Edge Integrity fibers/cm.gm (knifecut)	4
	Cut Edge Curl	0.5 cm
	Surface Interlocking	0.1 gm
	Pucker Factor	4
	Porosity	0.5 K Pa.s/m

<u>GUARANTEED PERFORMANCE</u>:

Cycle Time:	20 Seconds Maximum
Seam Quality:	Pucker - 0.1mm/mm
Pickup Reliability:	99.5% Minimum
Accuracy:	+/- 1/8" Maximum

(Fig. H)

LIMP MATERIALS RESEARCH AT NORTH CAROLINA STATE UNIVERSITY

T. G. Clapp and D. R. Buchanan
Department of Textile Engineering, Chemistry, & Science
College of Textiles
North Carolina State University
Raleigh, NC 27695-8301

NATO ASI Series, Vol. F 64
Sensory Robotics for the Handling of Limp Materials
Edited by P. M. Taylor

INTRODUCTION

The College of Textiles at North Carolina State University is committed to support fundamental and applied research to enhance international competitiveness of the U. S. fiber, textile, and apparel complex. Automation of apparel manufacturing is an important part of the overall research in this area. Our research programs in apparel automation have some unique characteristics, particularly in our emphasis on the need for quite sophisticated materials knowledge in the design and control of appropriate, flexible automated processes. At the same time, our research programs have benefited greatly from close interactions with the major industry-sponsored research programs in this area: those at Textile/Clothing Technology Corporation [$(TC)^2$], The Charles Stark Draper Laboratories, and the Singer Sewing Company.

Successful and general automation of the handling of material requires an understanding of the material behavior during the handling process. Unlike most conventional rigid materials, limp materials, particularly apparel fabrics, have inconsistent and highly variable behavior under load. This behavior is influenced by a number of factors, such as variations in external loading conditions, environmental conditions, and variation of material properties caused by the inherent variability of fibers, yarns, fabrics, and processing conditions. In addition, predicting limp fabric behavior is limited because of the limited applicability of current theoretical models.

Our fundamental philosophical approach to defining research programs in the area of automated handling of limp materials is to identify governing material parameters influencing the handling process and to modify the material or the handling process to minimize or eliminate the effect of material property variation. Research in techniques and models for predicting material behavior given the governing material parameters is also a part of this approach, as is the realization that automation of limp fabric handling requires sensor technology to measure governing parameters and use this information to control the handling process.

LIMP FABRICS RESEARCH PROGRAMS

At North Carolina State University, several research projects address roadblocks preventing the automated handling and high speed sewing of apparel fabrics. Among them are the following:

1) Optimization of a Flexible Ply Separation System for Feeding Small Parts - T. G. Clapp

The Defense Logistics Agency is sponsoring a major apparel automation program in the College of Textiles at North Carolina State University. The goal of this program is to develop a flexible, affordable, and modular work cell for the assembly of apparel products. Four basic modules are being developed to have the flexibility to work in various combinations to load, position, move material under the sewing head, and dispose of the joined parts. At this point in the research, the feeder module is near completion.

The purpose of this particular project is to identify the material parameters that influence the successful operation of the ply separation device in the feeder module. To accomplish this task, fabric material properties must be determined and related to the performance of the handling system using selected fabrics at various mechanical parameter settings.

The JetSew Clupicker has been selected as the ply separation device to be used in the feeder module. The Clupicker [1], which is illustrated in Figure 1, represents a production-proven technology, and its use therefore minimizes development risks and costs. Two mechanical parameters, the gap width and the contact force, can be adjusted on the Clupicker when material properties vary from style to style.

The Kawabata Evaluation System is used to characterize the material properties of a variety of military apparel fabrics. The important material properties include: thickness, bending rigidity, surface roughness, surface friction, tensile modulus, shear modulus, and compressibility.

The mechanical performance of the Clupicker is determined by conducting ply separation tests for each fabric type under selected mechanical conditions. The experimental system, which assures repeatable, accurately controlled mechanical parameters during ply separation tests, consists of an IBM 7545 robot with a force-torque transducer on the robot wrist and the Clupicker as the end effector, as shown in Figure 2. The robot is programmed to apply a selected compressive load on a fabric stack and control the pickup operation. During each test the mechanical system must accurately perform the sequence of ply separation operations. Analysis of these data determines the mechanical performance of the Clupicker relative to selected mechanical parameters and fabric type.

Preliminary results from this study show that the material parameters with the greatest influence on mechanical performance are thickness, bending rigidity, and surface roughness. These results support the results of a similar study conducted by Brotherton and Tyler [2]. The second phase of the project will be focused on developing technologies to measure these governing parameters as close to the handling operation as possible and to use this information to optimize the mechanical parameters. In the future, the goal is to develop automated handling systems that can measure the governing material parameters and automatically adjust mechanical parameters to optimize performance. In addition, low cost sensor technology is needed to measure fabric material parameters and control automated handling devices.

2) Models of Large Deflection Bending of Woven Fabric for Automated Material Handling - P. R. Brown, T. G. Clapp, D. R. Buchanan

The purpose of this project is to determine the engineering principles needed to design automated handling systems that are responsive to differences in fabric deformation characteristics. A methodology has been developed for predicting the optimum path for the robotic laying or picking up of apparel fabrics. The research goals are to 1) predict fabric deformation, given physical property information, as it is being placed onto a work table or removed from a work surface and 2) use the deformation information to determine the robot trajectory, as illustrated in Figure 3.

Classical theories for describing fabric deformation have been developed by Pierce [3], refined by Grosberg [4], and solved numerically by Konopasek [5], Lloyd [6], and Ghosh [7]. But, very limited experimentation has been conducted to validate these models, especially with respect to their applicability to predicting the behavior of apparel fabrics.

Based on the formulations used by previous researchers in this field, a mathematical model was developed to predict fabric deformation (Figure 3 [8]). The physical properties of length, width, bending rigidity, and weight per area define the fabric. The model computes the fabric deformation at discrete points as the fabric is laid onto the table. The usefulness of the model was evaluated by comparing the predicted fabric deformations with actual fabric deformations. Three apparel fabrics with different physical properties were analyzed.

The experimental system for measuring the actual deformed fabric shape consisted of an American Robot with six degree of freedom and an IRI image processing system (Figure 4). The robot moves the fabric sample to selected locations, and the vision system captures the image of the deformed fabric shape. The image is then scaled and compared with the predicted deformed shape (Figure 5).

The results of this study validate the applicability of the model to predict apparel fabric deformation for lengths in the 10 cm range. Although refinements are needed to improve the model, this study provides an approach by which knowledge of the physical and geometrical properties of the fabric parts can be used in a computer model to compute the trajectory of a fabric piece in an automated handling system. Additional research is needed to improve the model to include general geometrical features of the fabric parts and improve the computational efficiency of the algorithms.

3) Using Material Behavior to Optimize Ply Separation in an Industrial Application - T. G. Clapp

The purpose of this project is to improve the mechanical efficiency of an apparel fabric ply separation system for a specific apparel company. While the specifics of the research are proprietary, the methodology for solving the industrial problem is the same methodology used in fundamental research. The strategy is to analyze the properties of the material, identify those properties influencing the operation of the handling system, and modifying the material or the mechanical system to optimize mechanical performance. The first phase of the investigation involved a study of the ply separation system during production. A Parato chart was constructed to rank causes of handling problems. The two major causes in this particular application are 1) picking an incorrect number of plies and 2) distortion of the second ply owing to excessive attractive forces between plies.

In the next phase, a number of fabric styles with good and poor performance histories were characterized by measuring bending rigidity, surface roughness, surface friction, weight per unit area, thickness, and fabric-to-fabric friction. These fabric properties were related to mechanical performance. The results of this analysis show that thickness, surface roughness, and bending rigidity are greatly affected by one of the manufacturing processes. Thickness is also highly variable between lots of the same style.

Since uniform thickness is a key parameter for good performance in an automated handling system, an additional study of fabric thickness was conducted. The thickness of different fabric lots within the same style was measured under various compressive loading conditions. The results of this study show that as the compressive loading is increased, the variation in thickness between lots and due to variations in compressive loading decreases (Figure 6).

Based on this study, several recommendations have been made to improve the mechanical performance of the handling system, and one of the preceding manufacturing processes was identified for improvement. The fundamental problem with that particular manufacturing process is that no sensor technology is applied to monitor and control it.

The variability of one of the critical parameters, fabric thickness, can be reduced by increasing the compressive loading on the fabric stack. And since fabric thickness is so critical for proper handling, the initial stack thickness is measured and related to a mechanical setting on the ply separation device to enable the operator to optimize that setting.

Understanding the influence of material behavior on automated handling enables this apparel company to address the most significant problems preventing high mechanical efficiency. Further application of sensor technology is needed to monitor and control the handling process; but equally important, there is a need to expand these applications to all manufacturing processes.

4) A Sewing Dyomometor for the Study of High Speed Sewing - A. Matthews, T. J. Little

Sewing is the primary method for joining apparel products. Sewing speeds are increasing, and automated work cells are being developed. The interaction between limp fabric properties and behavior and the mechanical handling and joining systems are also changing and must be understood. The purpose of this project is to develop a system to monitor high speed sewing and provide a deeper understanding of needle penetration and fabric feeding forces.

The NCSU Sewing Dynamometer is constructed by instrumenting a microprocessor controlled sewing machine with piezoelectric force transducers mounted into the shafts of the needle bar and the presser bar [9]. The force signals are recorded simultaneously on a disk recorder from a two channel digital oscilloscope (Figure 7).

The Sewing Dynamometer is used to measure the needle forces encountered in the sewing of various fabrics at high speeds. An example of a typical force trace is shown in Figure 8, in which critical points in the sewing cycle are identified. The magnitude of the force at these critical points varies with fabric construction and with the number of plies being sewn. The force information from this study expands the understanding of the actual forces generated during high speed sewing with a variety of fabrics. This information can be used to predict high speed sewability of apparel fabrics.

Additional sensor technology will be added to the Sewing Dynomometer to detect quality problems during high speed sewing, such as seam puckering and errors in stitch formation. The future of flexible, high speed sewing may depend on the ability of the joining system to continuously monitor and control mechanical parameters.

5) Programs of The Consortium for Research in Apparel, Fiber & Textile Manufacturing (CRAFTM) - P. Banks-Lee, R. L. Barker, D. R. Buchanan, T. G. Clapp, J. W. Eischen, T. K. Ghosh, B. S. Gupta, H. Hamouda

The Consortium for Research in Apparel, Fiber & Textile Manufacturing (CRAFTM) is sponsored jointly by industry, the College of Textiles, and the State of North Carolina. Its purpose is to conduct interdisciplinary research programs with the goal of enhancing the international competitiveness of the U. S. fiber, textile, and apparel industries. CRAFTM also will be heavily involved in the transfer of research results to the industry, and in the education of students. Industry involvement at every stage of the program is a central feature.

Specific initial research areas include: developing system dynamics models to give U. S. manufacturers the capability to respond rapidly to market needs; relating critical textile material properties to the requirements of automated manufacturing; and improving batch wet processing technology to minimize product variability. The second of these research areas relates directly to the limp fabric research described above. It has the general title of: Dynamic Interactions of Fabrics with Manufacturing Processes.

Progress toward modernizing apparel manufacturing processes is hindered by the lack of a thorough understanding of the interactions between fabrics and those manufacturing processes. Several major roadblocks can be identified, each of which concern either the control of automatic limp fabric handling, optimization of assembly performance, or the assurance of product quality through elimination of manufacturing defects. This research program is divided into four sections:

1. Computer Simulation of Flexible Fabric Motion. This project addresses problems associated with the transport of flexible fabrics, using advanced mechanical analysis and sophisticated computer techniques to simulate the motion of fabrics as they interact with textile and apparel manufacturing machinery. Such computer simulation capabilities have significant industrial applications; they will provide designers and/or operators of fabric handling equipment with the means of making systematic machine adjustments to optimize the performance of fabrics with different material characteristics in the system.

2. Influence of Fabric Friction on Automated Handling and Transport. While a better understanding of the role of fabric friction as a predictor of fabric response during automated handling and high-speed assembly operations is fundamentally useful, it has specific utility as it contributes to the solution of the classic problem associated with textile fabric handling: removing single layers of fabric from the top of a stack in a reliable manner.

3. Developing Methods for Measuring Fabric Shear Properties Critical to the Control of Garment Assembly Operations. Fabric shear properties are known to be sensitive indicators of handling behavior and drapeability. Data on the pure shear properties of fabrics are also essential for certain computational models that predict fabric handling and/or drape from a knowledge of mechanical properties. Reliable and repeatable methods for measuring shear properties of fabrics are missing, and need to be developed.

4. Material Properties and Seam Puckering Problems in Sewing. This project targets the relationship between material properties and manufacturing problems affecting the acceptability of finished clothing. The research will identify those factors contributing to the development of seam puckering in sewn fabrics. The Kawabata Evaluation System (KES) will be used to determine fabric mechanical properties and their correlation with seam puckering under controlled sewing conditions. This research will provide the basis of a sewing control system that ultimately can be used as a systematic process for making machine adjustments to minimize sewing defects, and is particularly relevant in assembly systems with fewer and fewer human operators.

6) Programs of Textile/Clothing Technology Corporation [(TC)^2], The Charles Stark Draper Laboratories, and The Singer Sewing Company

With last year's establishment of the (TC)^2 Technology Center in Raleigh, a new era in industry-university relationships was entered. While the

Technology Center and North Carolina State University are separate organizations, they mutually enjoy an interactive relationship that is of benefit to both parties. While the purpose of the university's research activities is to perform fundamental research, some applied research, and to educate students, that of the Technology Center is "to encourage apparel manufacturers to take advantage of existing technology" [10]. However the technology it is showing includes the vision-guided robotic sewing systems developed at Draper Laboratories under $(TC)^2$ sponsorship.

Part of the reason for locating the Center in Raleigh was to afford the opportunity to interact with university programs. The first research project sponsored by the Center involves Industrial Engineering faculty and students, and is directed towards the comparison of three advanced computer simulation and modeling systems, and their applicability to apparel industry requirements. One or more of these systems is expected to be able to compare equipment, reduce new system evaluation risk factors, evaluate product or equipment changes, and evaluate scheduling and other operational factors.

The centerpiece of $(TC)^2$ technology, however, is the "robotic sewing machine" developed at Draper Laboratories and now being converted to a commercial version by Singer Sewing Co. Briefly, the concept can be described as follows. For the version as originally designed for assembling the two halves of a man's coat sleeve, the top sleeve and the undersleeve panels are aligned by the operator and layered on the belts of the autoloader. They are then transported to a vacuum table, where a vision system identifies their location, determines their size, and whether they are left or right sleeves. At this time, the exact position of the sleeve on the table is determined, and the mathematical expression defining the exact inseam to be sewn is established. The parts are transferred onto a differential belt system that feeds the parts into the sewing module under computer control. There the sewing head, also under computer control, is driven in a direction perpendicular to the controlled motion of the parts. This mechanism allows any seam contour to be sewn. Once the inseam is sewn, the sleeve is driven back to the vacuum table where the vent fold is made by a robot, under machine vision control. The partially finished sleeve is returned to the sewing module where the fold is tacked in place and then is brought back to the vacuum table where the parts are positioned for sewing of the outseam. The outseam is then sewn and the finished sleeve is ejected from the system [11]. This device sews sleeves at a rate of about four times that of a human operator.

While coat sleeves were the developmental target, this technology is quite general. Adaptations of it to the assembly of pants inseams and coatback seams, where controlled sewing of the very long seam is an advantage, are far along in development, while an adaptation to accept knit fabrics is presently being developed. The Singer modifications of this concept include breaking the original into interchangeable modules, to make it easier to adapt this technology to a multitude of sewing tasks.

A Singer technology already available to the industry is actually a family of automated systems. This family is called MARS (Manufacturer Applied Robotic Systems), each member of which utilizes one or more gantry robots with up to four degrees of freedom. These systems can be configured in various sizes and can be equipped with various sewing, feeding, cutting, marking and stacking devices to handle a wide range of applications. They had their origin, not in the apparel industry, but rather in other segments of the sewn products industry, particularly for automotive sewn products [11]. The concepts have progressed to the point where Singer now offers a jeans assembly system that combines four sewing heads, four loading stations, two robots and two folders in a single unit that provides fully automated assembly and bagging of a front pocket. MARS machines are viewed as complimentary technologies to the $(TC)^2$/Draper devices, with both essential elements of an automated apparel plant. The MARS machines represent, in general, less automation flexibility, and are really designed for small parts production. The $(TC)^2$/Draper/Singer machine, on the other hand, represents maximum automation flexibility, largely because of its use of vision systems, computer-controlled sewing heads, and flexible material transfer. It is viewed as the system of choice for working for long seams.

CONCLUSIONS

A substantial number of research programs in the handling and assembly of limp fabrics are being carried out at North Carolina State University. In the College of Textiles, these programs are funded both by industry and state sources, as well as from College funds directly. Close relationships with the major industrial participants in automated apparel manipulation and assembly research has proven highly beneficial for all participants.

REFERENCES

1. Hall, M.K., Bill Blackwood, and Bill Loweth, "Methods for Auto-Separation of Fabric Plies From a Cut Stack," Knitting International, September1987, pp. 162-166.
2. Brotherton, Allen A. and David J. Tyler, "Clupicker Performance and Flexible Apparel Automation," Hollings Apparel Industry Review, 1983, 3(2), pp. 15-34.
3. Peirce, F.T. IV-The Geometry of Cloth Structure. Shirley Institute Memoirs. Vol. XV, 1936, pp. 65-116.
4. Grosberg, P. "The Mechanical Properties of Woven Fabrics Part II: The Bending of Woven Fabrics," Textile Research Journal, March, 1966, p. 323-337.
5. Konopasek, Milos. "Computational Aspects of Large Deflection Analysis of Slender Bodies," in Proceedings of the NATO Advanced Study Institute on Mechanics of Flexible Fibre Assemblies, The Netherlands, 1978.
6. Lloyd, D. W. and W. J. Shanahan, "The Folding of Heavy Fabric Sheets." Int. J. Mech. Sci. Vol. 20. pp. 521-527, 1978.
7. Ghosh, Tushar K., "Computational Model of the Bending Behavior of Plain Woven Fabrics," Ph. D. Thesis, College of Textiles, North Carolina State University, 1987.
8. Brown, P. R., "Large Deflection Bending of Woven Fabric for Automated Material Handling," Master's Thesis, College of Textiles, North Carolina State University, Raleigh, NC, 1988.
9. Matthews, Ann, "Development and Evaluation of the NCSU Sewing Dynamometer for the Determination of Fabric Sewability," Master's Thesis, College of Textiles, North Carolina State University, Raleigh, NC, 1987.
10. Schaffner, Karen, "Technology on Trial", Apparel Industry Magazine, p. 28, April 1988.
11. Vento, Vince, "Applying $(TC)^2$ and MARS Technology to Automation of the Sewing Process", Paper delivered at the 7th International Technical Meeting and National Convention of the Clothing & Knitting Industry, Milan, 13 June 1987.

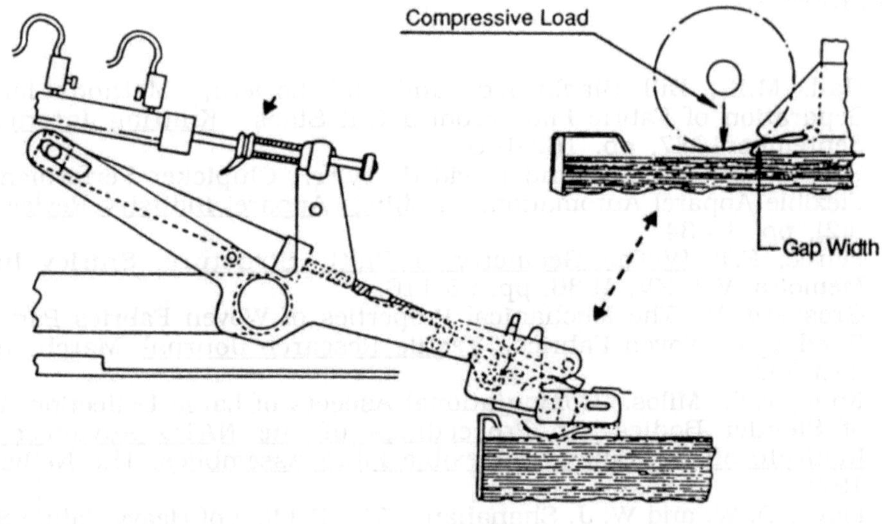

Figure 1. The Clupicker with adjustable mechanical parameters highlighted.

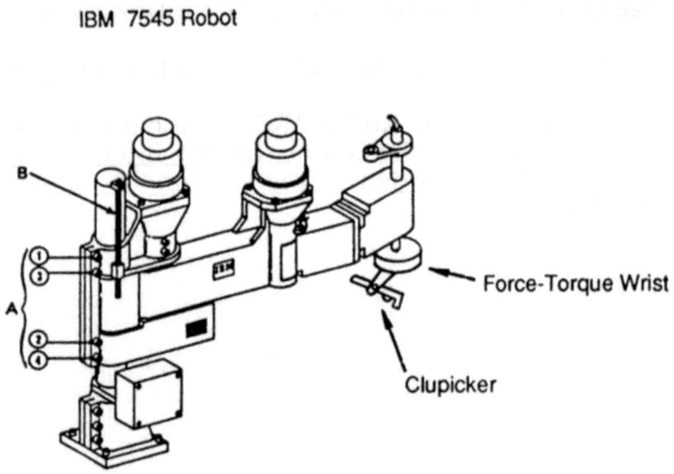

Figure 2. Experimental system for testing ply separation.

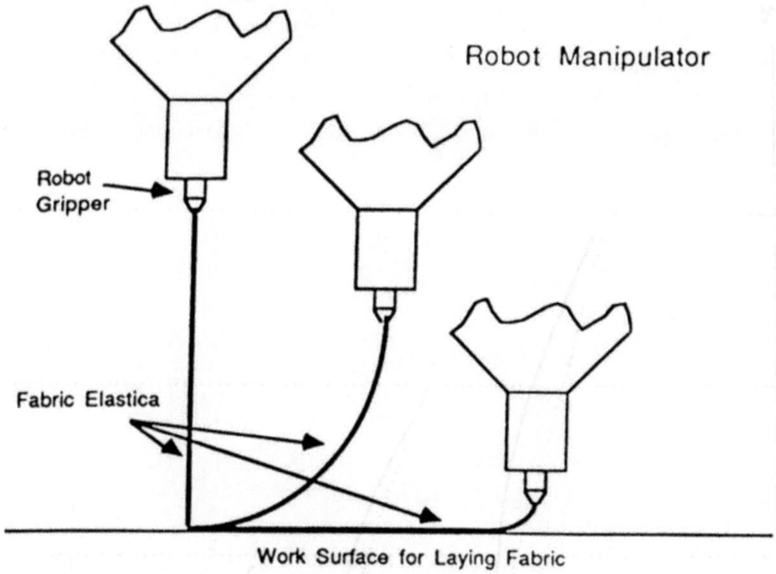

Figure 3. Laying of fabric onto a work surface.

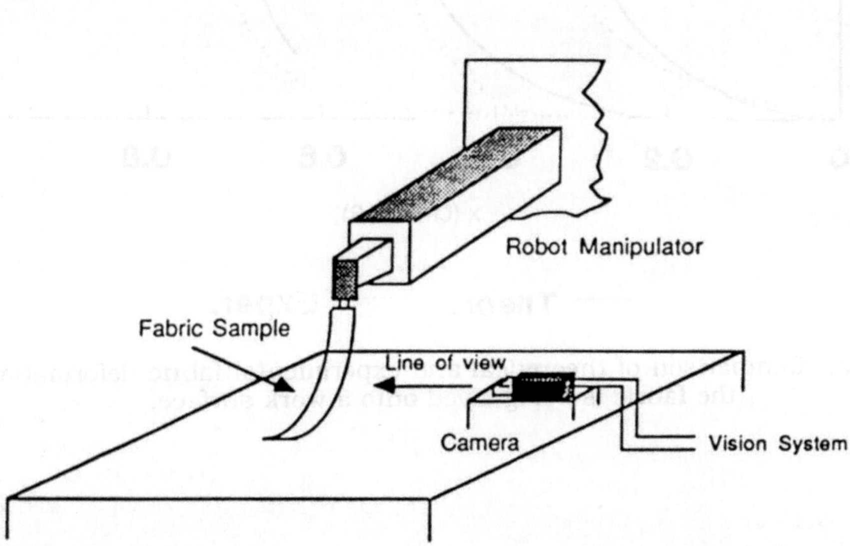

Figure 4. Experimental apparatus for measuring fabric deformation.

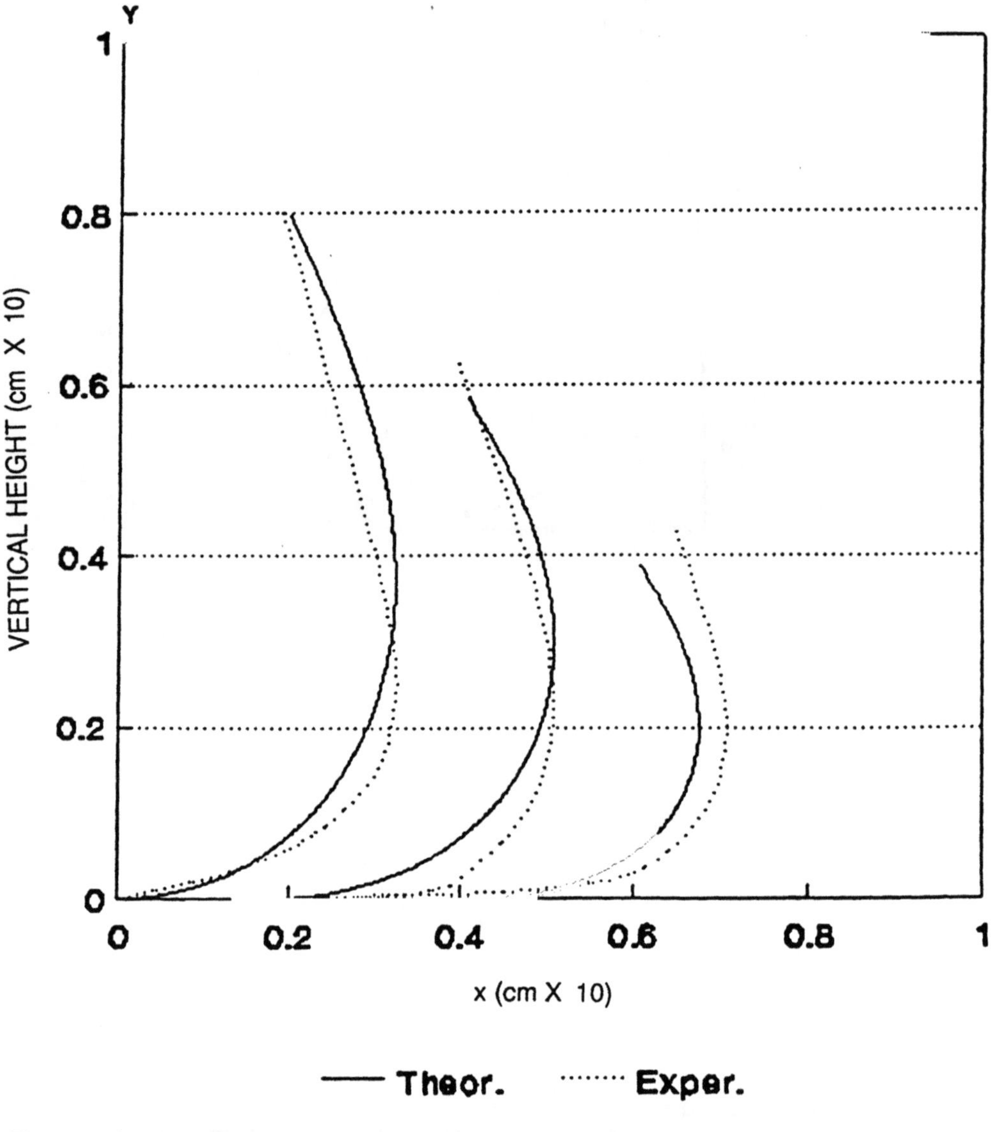

Figure 5. Comparison of theoretical and experimental fabric deformation as the fabric is being layed onto a work surface.

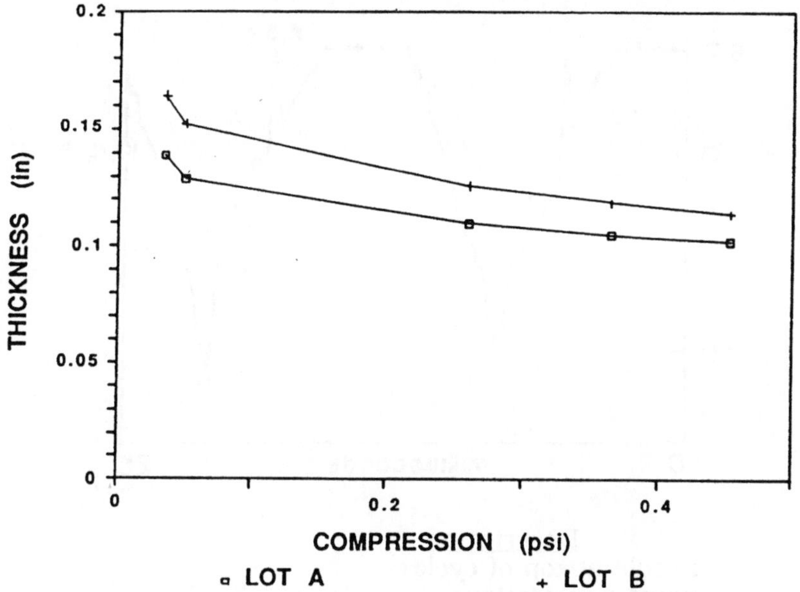

Figure 6. Effect of compression on fabric thickness using two lots of the same style of fabric.

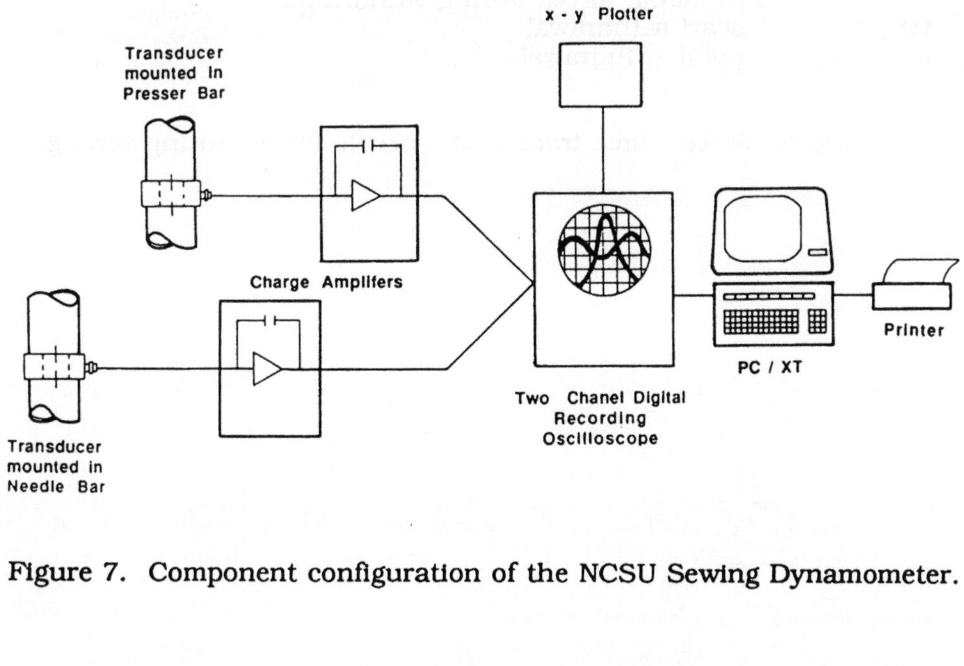

Figure 7. Component configuration of the NCSU Sewing Dynamometer.

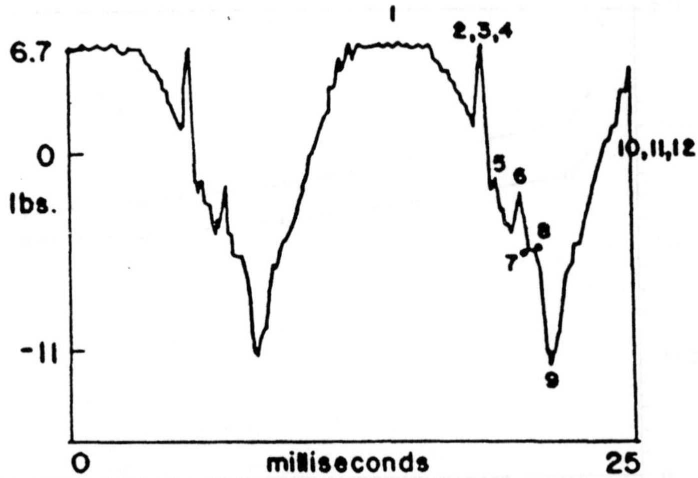

Point	Description
1	needle at top of cycle
2	point penetration
3	eye penetration
4	scarf penetration
5	frictional forces on the shank of the needle
6	shoulder of the needle penetrates
7	lowest position in the cycle
8	marks the change in inertia, by the needle bar
9	frictional forces during withdrawal
10	scarf withdrawal
11	point withdrawal

Figure 8. Force - time trace of the needle forces during sewing.

SENSORY ROBOTICS: IDENTIFYING SEWING PROBLEMS

S.C. Harlock, D.W. Lloyd
G. Stylios *
Department of Textile Industries
University of Leeds
Leeds, LS2 9JT
England

ABSTRACT

An essential feature of any automated garment manufacturing system is that it produces high quality seams, i.e. seams free from pucker, free from damage, with properly balanced sewing threads, etc. If poor quality seams are produced, all the sophistication of automated production will be wasted, as the garments produced will be commercially unacceptable.
The various types of seam faults may be identified in different ways; this has consequences for the way in which sensing is carried out. Some "sensing", because it relates to fabric or thread properties, is most appropriately carried out as part of quality control on incoming materials and not at the point of production. Other problems, because they arise as part of the sewing operation, are best identified at the sewing station.

Introduction

Automating manufacturing operations within the mass-production garment industry is commonly regarded as a route to greater competitiveness, through reduced labour costs, increased productivity, fewer rejects, reduced response times and reduced work-in-progress. Up to a point this is probably true, but the concentration on automating individual manufacturing operations overlooks the fact that making use of automated processes - islands of automation - or of a fully automated plant, demands new approaches to management and information handling. These techniques are being introduced into other industries further down the automation road than apparel, under the general title of Computer Integrated Manufacture, CIM. Although proposals for the development of CIM in garment production have been made [1], the concept of CIM is only slowly being understood in the clothing context.
An essential part of the new management approach to garment manufacture demanded by automation is an organisation-wide dedication to the highest

---

*School of Industrial Technology, University of Bradford, Bradford, BD7 1DP.

NATO ASI Series, Vol. F 64
Sensory Robotics for the Handling of Limp Materials
Edited by P. M. Taylor

quality standards. Consumers are becoming ever more quality conscious and demanding; the success of some imported Japanese fabrics may be seen as evidence of the dedication of that industry to quality and consumer preference (see, for example [2]). It is vital, therefore, that automation leads to increasing quality; being the most productive manufacturer of unsaleable garments has nothing to do with competitiveness.

Clearly, some causes of commercial unacceptability have little to do with automation, poor design and aesthetics being obvious cases. Some causes are indirectly linked. Marginally acceptable fabrics which are difficult to tailor may, with experienced personnel, be turned into saleable garments, but will not be processable by an automated system. In this context, automated manufacture will demand much more rigorous and well-defined quality control of incoming materials. Some causes of rejects are directly linked to automation; defects such as pucker and sewing damage occur in normal sewing operations [3]. In normal sewing, the machine operator should observe the problem and initiate remedial action, whereas in automated sewing such problems might pan undetected while a large number of defective items are produced. Finally, there are problems that arise with automated machines that should not occur with an experienced operative. For example, an experienced operative should anticipate exhaustion of the bobbin thread and take remedial action.

Devising sensing techniques to detect broken or missing threads, or other machine fault conditions, is part of the process of machine development and is not the concern of this paper. The more difficult sensing task is to detect fault conditions arising from interactions between the fabric and the sewing process [4]. Such fault conditions may be grouped under two headings according to the visual effect of the fault, seam pucker and seam damage. These two groups of faults will be considered in turn.

SEAM PUCKER

Seam pucker is defined as [5]: "a ridge, wrinkle or corrugation of the material or a number of small wrinkles running across and into one another, which appear when sewing together two pieces of cloth".

It is possible to identify five causes of pucker [6]:

(a) differential feed pucker
(b) thread tension pucker
(c) pattern piece mismatch
(d) laundering
(e) structural jamming pucker.

(a) Differential Feed Pucker

This type of pucker arises from the way that fabric plies are fed through the sewing machine. The lower ply is moved forward by the action of the feed dogs. The upper ply is moved forward by the lower fabric through the friction between the plies. The top ply is also subject to drag from friction with the presser foot. The upper ply may become stretched relative to the lower ply, with the result that after seaming the top ply tends to contract, compressing the lower ply along the seam and causing pucker.

Elimination of differential feed pucker requires attention to the way sewing is carried out. If different fabrics are to be joined, the relatively stiffer fabric should be placed on top. Slow sewing speeds, careful fabric handling and low presser foot pressures, or the adoption of positive feed devices are required. Since eliminating differential feed pucker requires alterations to the sewing operation, sensing during sewing is too late; the problem must be identified before production begins. "Sensing" in this case becomes a natural part of material selection and quality control on incoming materials.

Determining whether differential feed pucker is likely to cause problems with a particular combination of fabrics requires sample seams to be sewn under conditions as close as possible to actual production conditions. The authors are unaware of any work to correlate the tendency of a fabric to exhibit differential feed pucker with objective measurements of fabric physical properties, though there would appear to be potential benefits from such an exercise if surface friction were included in the properties measured. Discussion of appropriate instrumentation is deferred until later in the paper.

(b) Thread Tension Pucker

Even in the simplest types of seam, the needle and bobbin threads are in tension during the sewing operation. The needle thread tension is necessarily slightly greater than the bobbin thread tension, as it has to pull the bobbin thread into the fabric against the bobbin thread tension and the frictional forces between the threads and the fabric. If the threads are relatively extensible and recover as the seam is released from the presser foot, the contracting threads compress the fabric along the seam and pucker results.

Thread tension pucker may be identified by sewing a test seam with a layer of paper on top of the fabric. If, when the paper is torn away gently, the seam becomes pucker free, thread tension pucker is occurring. Elimination of this type of pucker requires the use of minimum thread tensions consistent with the production of a balanced seam [7]; if this is insufficient to eliminate the pucker, a change of sewing thread is required to a thread with minimum extensibility. This suggests that two types of sensing are required. The first is again a natural part of quality control, involving sewing test seams using the fabrics, sewing threads and sewing conditions (especially thread tensions) used in production. The second is to ensure that production equipment is set up with minimum thread tensions, and that these tensions are maintained, so that the choice of threads obtained from quality control on the incoming materials remains relevant to production. Clearly, there is no need for the computer system controlling automated production to be burdened with monitoring all thread tensions continuously. A more appropriate technique would be to use low cost, dedicated devices attached to each needle thread-line, to flag an error condition electronically when thread tensions exceed preset limits for more than a certain proportion of the sewing cycle.

(c) Pattern Piece Mismatch

During making-up, garment pieces must match for length along the edges of a seam, except where some mismatch is deliberately introduced to produce shaping of the garment. There is a limit to the relative difference in lengths beyond which compression of the longer edge by the stretched shorter edge produces pucker [8]. The appearance of this type of pucker is characteristic, allowing visual identification. The limit to the relative difference in lengths depends on the ability of the fabric to deform in plane without buckling. This, in turn, depends on the fabric mechanical properties and the orientation of the seam.

The length mismatch required is a function of individual garment design and is known in advance. The ability of a fabric to accommodate this mismatch, since it depends on fabric mechanical properties, can be predicted from measurements of fabric properties [9,10]. Therefore, such problems of tailorability also form a natural part of quality control on incoming materials using objective measurement techniques.

(d) Pucker Caused by Laundering

If the sewing thread is dimensionally unstable, so that the thread shrinks during washing, wetting, or other after-treatments, pucker will occur [11].

This is not strictly a sewing problem, but again is part of quality control applied to sewing threads. There is an interaction with other aspects of sewability, however, in that many factors influence the choice of sewing thread, so laundering pucker must be considered when sewing threads are being chosen if any wet processing is to take place.

(e) Structural Jamming Pucker

This type of pucker [3] occurs mainly in woven fabrics, usually plain weave, sett close to the maximum theoretical sett, i.e. fabrics with nearly as many ends and picks per centimetre as it is practically possible to weave. The insertion of sewing thread into a fabric to form a stitch requires a local readjustment of the fabric structure to make room for the extra threads. In fabrics prone to structural jamming pucker there is insufficient space in the structure for local movements of the yarns to take place easily; the resultant squeezing of the fabric structure between stitches causes buckling which appears as seam pucker [12].

Structural jamming pucker is independent of sewing machine mechanics, so it is vital to be able to diagnose this type of pucker correctly, otherwise considerable time and effort will be expended in making unnecessary and ineffective sewing machine adjustments. This type of pucker may be identified by cutting the stitches of a test seam on both sides of the fabric, leaving the threads penetrating the fabric. If pucker remains after the stitches are cut, the cause is structural jamming. Structural jamming pucker may be eliminated by modification to the design of the garment, to ensure that seams are aligned at a slight angle to the warp or weft directions, rather than coincident with them, to benefit from greater fabric compressibility [12], changing the number of stitches per cm, reducing thread thickness, reduction of warp and weft cover factors [13], or locally stiffening the seam line [6].

It follows that since structural jamming pucker arises from the properties of the fabric, it is possible to predict it from measurements of fabric properties [6,14]. Again, "sensing" for this type of pucker becomes part of quality control on incoming materials, either through the production of test seams, or through objective measurement of fabric properties.

Other Contributory Causes of Pucker

The mechanisms described above are all static in nature. However, at modern sewing speeds dynamic conditions exist at the sewing point. It can be shown [14] that presser-foot bounce causes a loss of control of both the fabric and sewing threads at a critical time, and hence contributes to seam pucker. It is possible to monitor presser-foot bounce and to eliminate it by providing suitable means of damping the system and reducing presser-foot mass; this has a beneficial effect in reducing pucker. Eliminating presser-foot bounce at achievable sewing speeds is strictly the province of the sewing machine makers, but if severe presser-foot bounce is a recurrent problem, its onset could be monitored by a suitable displacement transducer.

SEWING DAMAGE

Sewing damage is the formation of holes in a fabric during needle penetration by severance of one or more yarns. Two basic causes of sewing damage can be identified [14]:

(a) Frictional heating of the needle and yarn melting
(b) Mechanical breakage of yarns.

(a) Frictional Heating of the Needle and Yarn Melting

During sewing the needle is subject to frictional heating from two sources; friction with the fabric layers during penetration, and friction with the needle thread as it passes through the eye. The needle is also subject to cooling, from its movement through the air, from movement of air through the eye entrained by the thread, and from conduction of heat into the thread and any lubricants on it. Depending on the relative importance of the various heating and cooling mechanisms at the relevant sewing machine speeds, the needle, or at least that part of it which penetrates the fabric, will reach a new equilibrium temperature very quickly.

If heating mechanisms dominate, the needle may reach temperatures high enough to melt synthetic filaments. The result is a series of holes in the fabric marking each needle penetration, where the yarns around the edges of the hole are melted [15]. The resulting seam is at best unsightly; in some knitted fabrics the structural integrity is lost, so that "ladders" are produced.

The solution to such thermal damage is to reduce needle heating and increase needle cooling so that a lower needle equilibrium temperature is achieved. Thread and fabric lubricants reduce friction with the needle and so reduce needle heating rates. Modern thread lubricants based on silicones are additionally capable of high rates of heat removal from the

needle, and direct needle cooling may be introduced in the form of air-jets or felt pads containing lubricant [16].
Direct sensing for high needle temperatures in an industrial context would be extremely difficult. In the past, needle temperatures have been monitored using thermocouples attached to the needle [17], or by pyrometric techniques [18], but only in a laboratory setting. Thermocouples interfere with normal sewing and would not survive for long in an industrial environment. Infra-red measurements would be swamped by background sources and the sensors unsighted by real production methods. It is also not possible to mount thermocouples or other devices high on the needle. The high thermal capacity of the needle bar and relatively poor thermal conductivity of the needle result in high temperature gradients above the heated zone, so that the rest of the needle is comparatively cold [16,18]. So although sensing of needle temperatures would provide a method of flagging the onset of likely thermal damage, in practice it is unlikely that such direct sensing could be arranged.
It is, therefore, important that sewing tests under production conditions be carried out in advance of production, to establish if a thermal damage problem exists. Care must be taken over thread lubricants and needle cooling devices where a problem of thermal damage is found to exist.

(b) Mechanical Breakage of Yarns

It is the job of the needle when sewing fabrics, in contrast to continuous materials, to form a hole for the threads by pushing the fabric yarns apart, not by cutting a hole. If yarns are severed by the needle, oversize holes are formed with the same associated problems as holes caused by thermal damage. Several factors combine to produce mechanical damage, which occurs most often in knitted fabrics during overlocking. If the yarns are prevented by high frictional forces at interlacings from sliding past one another easily, they can only accommodate the needle through extension. If the length of the yarn held between adjacent interlacings is too small, or the needle diameter is too large, the extension required may be greater than the yarn can accommodate. The difficulty is aggravated by oversize holes in the needle plate which lead to "flagging" - the depression of the fabric locally into the needle hole, with related extra extension of yarn segments. Problems are also associated with the use of needles with exaggerated cross-sectional shapes with large effective diameters or needles with inappropriate point geometries [14].

Mechanical damage may be reduced or eliminated by using smaller diameter needles and correctly-sized needle holes and reduced by a suitable choice of point geometry. The only effective way of eliminating mechanical damage is through correct fabric finishing, where application of fabric softeners lubricates the yarn interlacings, reducing inter-yarn friction and allowing easy slippage.

It follows that test seams should be sewn ahead of production to determine if a problem exists with the fabric and needles to be employed. However, it also follows that assessment of inter-yarn friction might allow problems of mechanical damage in a fabric to be predicted; thus measurements of hysteresis in mechanical properties or of surface friction may allow the prediction of mechanical damage difficulties. This again is part of quality control on incoming materials.

However, application of softeners is not necessarily uniform, so there is still a place for real-time monitoring at the sewing point to ensure that mechanical damage is not occurring during production. If a yarn is not free to slip or extend, the movement of the sewing needle forces the yarn to extend to break, and must, therefore, provide the force necessary to break the yarn. Needle penetration forces, therefore, provide a sensitive test of the onset of mechanical damage; when these forces are low, no damage is taking place, but above a critical value damage is likely to occur [14]. As with thread tension pucker, a dedicated monitoring device is the obvious choice. This would electronically flag the onset of damage-producing conditions. Measurements of needle penetration forces may be achieved by arranging a suitably strain-gauged cantilever section of the needle plate, by a separate cantilever around the needle hole, or by putting load sensors between the needle and needle bar or on the needle bar itself. Such systems may be made robust enough to withstand an industrial environment [14].

This approach also provides a means for investigating the tendency of a fabric to suffer damage, and for determining the choice of needle sizes, point shapes, etc., that are required to reduce the problem to acceptable proportions. This requires a suitably instrumented sewing machine; the details of such a system and use are described elsewhere [14]. It may also be the case that such a system, either in real-time or during quality control, would be capable of detecting excessive needle-fabric friction, and hence of detecting indirectly the possible onset of thermal damage.

INSTRUMENTATION FOR PREDICTING SEWING PROBLEMS

Measurements of fabric properties, especially mechanical properties, in the context of predicting which fabrics will tailor well and which badly, or which will be prone to sewing problems is not a new idea [19,20,21]. In the past the approach has failed to gain widespread adoption because subjective assessment by experienced personnel was adequate and because suitable instrumentation was not available commercially. Interest in instrumental methods of objective measurement has been revived in recent years, however, by the work of Professor Kawabata and the Japanese Hand Evaluation and Standardisation Committee [22]. Acceptance of objective measurement methods has been helped by the loss of experienced subjective assessors and by the commercial availability of a set of purpose-designed instruments, the "Kawabata" or KES-F testers [23]. A remaining inhibiting factor is the very high cost of these instruments. Garment manufacturers have proved reluctant to construct attachments for tensile testing instruments to perform such measurements, a practice common in University departments for many years. Widespread adoption of instrumental techniques may have to wait for the availability of much lower cost instrumentation; a potentially interesting development in this context is the FAST system (Fabric Assurance by Simple Testing), a set of instruments under development at the CSIRO Division of Wool Technology, Ryde, New South Wales, though these instruments are not commercially available at present [24].

It is necessary, therefore, to limit discussion of objective instrumentation methods to the KES-F system. In the present context four instruments are relevant. These are a combined tensile and shear tester (KES-F1), a bending tester (KES-F2), a compression tester (KES-F3) and a surface tester (KES-F4). All the instruments are non-destructive, in the sense that all tests are performed at low stress levels on samples which are not destroyed, though they are too small to be of use in subsequent garment manufacture. The KES-F1 tensile and shear tester performs a complete shear cycle to a preset angle in both directions, followed by a tensile test to a preset tension level. Both tests produce hysteresis charts from which a number of parameters are recorded. The KES-F2 bending tester performs a cycle of bending from zero to plus and minus preset curvatures, finishing at zero curvature. Several parameters are recorded from the hysteresis chart obtained. The compression tester squeezes a single thickness of fabric between parallel surfaces, up to a preset pressure and back. Again, a hysteresis curve is obtained which is characterised by several parameters.

Finally, the KES-F4 surface tester functions in two modes, a surface friction mode and a surface roughness or texture mode. Parameters are used that quantify the friction and geometrical roughness [22].
These instruments have attracted considerable attention in the context of quantifying handle, tailorability and appearance as single numerical values [25,26]. This is not the place for a discussion of this approach. The relevance of these instruments (and any other instruments used to make an equivalent set of measurements), is that they are already being used to accept or reject incoming fabric and to predict problems of tailorability, at least in a number of Japanese high-quality tailoring organisations [10]. There are also reports of KES-F instruments being used for similar purposes outside Japan and of a similar set of locally-made instruments finding widespread application in China.
Application of the KES-F measurements in this way is normally through the medium of so-called "snake-charts" [22,26]. Here the ranges of possible values of each of the various parameters are drawn as horizontal scale lines arranged in a vertical stack on a single sheet of paper. The limits of acceptability of each parameter at each extreme are marked; these limits are joined by lines whose wavy appearance gives rise to the name. A fabric with any parameter values outside these limits is unacceptable as it stands. Other limits may be marked for particular parameters; these may indicate potential sewing problems or related appearance or dimensional stability problems. Inside all the limits is a zone which indicates that a fabric with parameter values inside this zone is completely acceptable as it is.
Knowledge of the exact values at which the various limits are to be drawn is increasingly regarded as a valuable commercial asset, as it represents the quantified experience of the particular tailoring organisation. Equally valuable are the sets of instructions that accompany the charts, specifying the remedial actions to be taken when one or more parameters are outside acceptable limits, or actions to overcome predicted tailoring difficulties. Typically these instructions relate to special precautions to be taken in handling and sewing or to refinishing operations. The formulation of these instructions requires real understanding of the tailoring problems and the specific effects of different finishing operations; such understanding comes from extensive trials and objective measurements.
This level of technology is best seen in the Japanese producers of high quality men's worsted suits [10,27].

Other "Sensing" Techniques

In garment manufacturing operations carried out manually, experienced operatives should monitor a range of parameters which reflect the health of the sewing and related operations. For instance, in operations such as overlocking, where the fabric edges are trimmed of excess fabric, the quantity of fabric removed should be kept to a minimum, otherwise the dimensions of the garment pieces will be incorrect. In automated sewing operations it would be appropriate to monitor the quantity of fabric removed, by weighing each trimming individually. Since the trimmings are already removed by vacuum, techniques developed in another context [28] for weighing fibre tufts might be used.

Such monitoring of miscellaneous parameters is a natural part of production monitoring, something already established in other areas of the textile industry [29] and finding application in apparel manufacture [30]. Production monitoring can be used to "sense" subtle problems, such as production machines running slower than planned, or individual production machines more prone to breakdown than normal. Such faults are sewing problems in that they affect the productivity of sewing directly rather than through a high level of rejects. Production monitoring and the related analysis of data, perhaps collected over many shifts, used to detect subtle production problems, is a management function and hence relates to computer-integrated-manufacture.

However, with the emphasis on shorter production runs, collection of meaningful data is difficult. This is ironic in that with such an emphasis the ability to predict problems and ensure correct machine settings is even more important to get it right first time.

REFERENCES

1. ULIS, Ltd. et al, submission to B.R.I.T.E. research program.
2. M. Nitta, "Introduction to Objective Measurement Data of Fabric Mechanical Properties to High-Quality Suit Production", in "Objective Specification of Fabric Quality, Mechanical Properties and Performance", edited by S. Kawabata, R. Postle, M. Niwa; The Textile Machinery Society of Japan, 1982, pp. 321-329.
3. L.M. Scott, "Some Problems Relating to Sewing", Journal of the Textile Institute, 42 (8), 1951, pp. 653-660.
4. "A Review of Seaming Problems", Clothing Institute Journal, 14, (1), 1966, pp. 35-45.
5. The Shorter Oxford Dictionary, Vol.II, Oxford Press, 3rd Edition, 1967, pp. 1614.
6. G. Stylios, "Seam Pucker and Structural Jamming in Woven Textiles", M.Sc. thesis, University of Leeds, 1983.

7. A.M. Bertoldi, "An Investigation into the Effects of Seaming Variables on Fabric Pucker", M.Phil. thesis, University of Leeds, 1973.
8. T.J. Mahar, R.C. Dhingra, R. Postle, "The Investigation and Objective Measurement of Fabric Mechanical and Physical Properties Relevant to Tailoring", in "Objective Specification of Fabric Quality, Mechanical Properties and Performance", edited by S. Kawabata, R. Postle,M. Niwa, The Textile Machinery Society of Japan, 1982, pp. 301-315.
9. K. Ito, "Process Control for Tailoring Based on Objective Data about Fabric Properties - Progress in the Last Year", in "Objective Evaluation of Apparel Fabrics", edited by R. Postle, S. Kawabata, M. Niwa; The Textile Machinery Society of Japan, 1983, pp. 89-96.
10. M. Nemura, "A High-Efficiency Tailoring System for Men's Suit Production - Latest Progress", in "Objective Measurement; Applications to Product Design and Process Control", edited by S. Kawabata, R. Postle, M. Niwa; The Textile Machinery Society of Japan, 1986, pp. 243-250.
11. J. & P. Coats, Ltd., "Seam Pucker", Thread Technology - Part 2, Technical Papers, September, 1976.
12. T. Townsend and N.H. Chamberlain, "More Light on Structural Seam Pucker in Lightweight Fabrics", Clothing Institute Technological Report No. 16, September, 1964.
13. T. Townsend, "The Factors Affecting Structural Seam Pucker in Man-made Fibre Fabrics", M.Sc. thesis, University of Leeds, 1964.
14. G. Stylios, Ph.D. thesis, University of Leeds, 1988.
15. Schmetz handbook.
16. Singer, Private Communication.
17. G. Stylios, Ph.D. thesis, University of Leeds, 1988.
18. Schmetz, Private Communication.
19. J. Lindberg, L. Waesterberg, R. Svenson, "Wool Fabrics as Garment Construction Materials", Journal of the Textile Institute, 51, 1960, T1475-1493.
20. J. Lindberg, B. Behre, B. Dahlberg, "Mechanical Properties of Textile Fabrics", Textile Research Journal, 31, 1961, pp. 99-122.
21. J. Westerfeld, F.J. Van Krugten, T.N.O. Report No. 13030 (1226-030) 13, 1967.
22. S. Kawabata, "The Development of the Objective Measurement of Fabric Handle", in "Objective Specification of Fabric Quality, Mechanical Properties and Performance", edited by S. Kawabata, R. Postle, M. Niwa; The Textile Machinery Society of Japan, 1982, pp. 31-59.
23. Product Technical Literature, Kata, Ltd.
24. Private Communication, May, 1988.
25. R. Postle, "Objective Evaluation of the Mechanical Properties and Performance of Fabrics and Clothing", in "Objective Evaluation of Apparel Fabrics", edited by R. Postle, S. Kawabata, M. Niwa; The Textile Machinery Society of Japan, 1983, pp. 1-14.
26. M. Niwa, S. Kawabata, K. Isizuka, "Recent Developments in Research Correlating Basic Fabric Mechanical Properties and the Appearance of Men's Suits", in "Objective Evaluation of Apparel Fabrics", edited by R. Postle, S. Kawabata, M. Niwa; The Textile Machinery Society of Japan, 1983, pp. 67-82.
27. K. Ito, S. Kawabata, "Conception of the Automated Tailoring Controlled by Fabric Objective-Measurement Data", in "Objective Measurement: Applications to Product Design and Process Control", edited by S. Kawabata, R. Postle, M. Niwa; The Textile Machinery Society of Japan, 1986, pp. 175-181.
28. Automatic fibre tuft strength tester, ATLAS technical information.
29. Dextralog, Ltd. Product information.
30. Eildon Electronics, Ltd. Technical information.

PROGRESS TOWARDS AUTOMATED GARMENT MANUFACTURE

P.M.Taylor, G.E.Taylor
Department of Electronic Engineering
University of Hull
Hull HU6 7RX, U.K.

KEYWORDS: Garment manufacture, limp materials, robotics, automation, electrostatics, vision feedback.

ABSTRACT

The handling tasks associated with automated garment manufacture are categorised into ply separation, pick and place, 2D manipulation and 3D manipulation. Progress in each of these areas is illustrated through the various garment manufacturing research projects at Hull University since 1981.

INTRODUCTION

The research projects undertaken at Hull in the area of garment assembly have all followed the same pattern. Particular problems are first identified with industrial collaborators. These problems have to be both relevant to the industrialists and must also contain a significant research content. This latter part makes the projects of interest to the University and allows approaches to be made to Government agencies which support research in the academic community. At the end of the research period of three years a laboratory prototype system is demonstrated at the University.

The first project, in collaboration with Corah plc and Marks and Spencer plc resulted in a robotic system which could separate plies of knitted cotton from a stack and place a motif in the correct position and orientation on the separated panel. The next stage was to investigate a garment subassembly: putting together the front, back and gusset panels of mens' and womens' underwear. Current work, with support from Corah plc and UMI Ltd, is concerned with the automated assembly of a complete underwear garment. In all of these projects, emphasis has been placed on combining flexible handling techniques with existing sewing machinery rather than developing hard automation. A further project has been carried out, in parallel with the above, on automated shirt collar inspection and assembly with support from I.J.Dewhirst plc, Marks and Spencer plc and assistance from the Reliant Tooling Company Ltd. All of these projects have received the major part of their funding from the ACME Directorate of the Science and Engineering Research Council.

NATO ASI Series, Vol. F 64
Sensory Robotics for the Handling of Limp Materials
Edited by P. M. Taylor

PLY SEPARATION

There is a long history of attempts at automated destacking mechanisms (Murray, 1975). Back in 1981, the only commercially available device was the Clupicker (Dlaboha, 1981), and in practice this was used for only denim and shirt materials. The requirement for the ply separation/motif application project was for a mechanism with >>99% reliability of single ply separation which could achieve separation (not necessarily removal) in 3 seconds or less. The device must work for a range of fabrics (the most important being knitted cottons), without delicate adjustments being necessary at change time. The device had to be non-intrusive to avoid fabric damage. This latter requirement obviated the use of pins.

Major factors which affect the separation process are the cohesive forces between layers and the degree to which edges are bound together, either by the edges fusing during cutting or because of interlocking fibres. The cohesive forces are particularly important when freshly cut stacks of knitted cotton are used since these are often compressed at the cutting stage. A further complication is that cutting is centralised in many companies and fabric stacks may be transported perhaps 100 miles to the making-up factory. The stacks will then no longer have perfectly vertical edges and this must be tolerated by the separation mechanism.

A new method was invented by Kemp (Kemp et al, 1982, 1986) using a sensory air jet/finger device shown schematically in figure 1. and attached to a Unimation PUMA robot. The device blows an air jet at an angle over the stack which induces vibrations in the top ply. In practice, because of the porosity of the fabric, some air penetrates the top ply, forming bubbles beneath it which are then propogated towards the edge of the stack. The combination of vibrations and moving bubbles breaks up the bond between the plies, and indeed does this particularly well at the edges where the most difficult bonding occurs. If the lower finger of the separation device is located just above the top edge of the stack, the vibrating top ply will flip over it , as seen in figure 2. This occurence can be detected by an infra-red sensor in one finger which detects the amount of light received from the emitter in the other finger. A high level indicates no fabric between the fingers, a medium level indicates one piece only and a low level means that there are two or more plies between the fingers. If one of the error conditions occurs the robot is programmed to retry the operation, failing this to relocate the top and edge of the stack and retry, and failing this to pause for operator assistance.

Once the edge of a single ply has been separated in this way, the fingers are slid under the topmost ply towards an edge where two infra-red emitter-detector pairs are used to accurately align the fingers with the edge before they are closed and the material is peeled away. This mechanism has since been further developed by Autotex plc as the

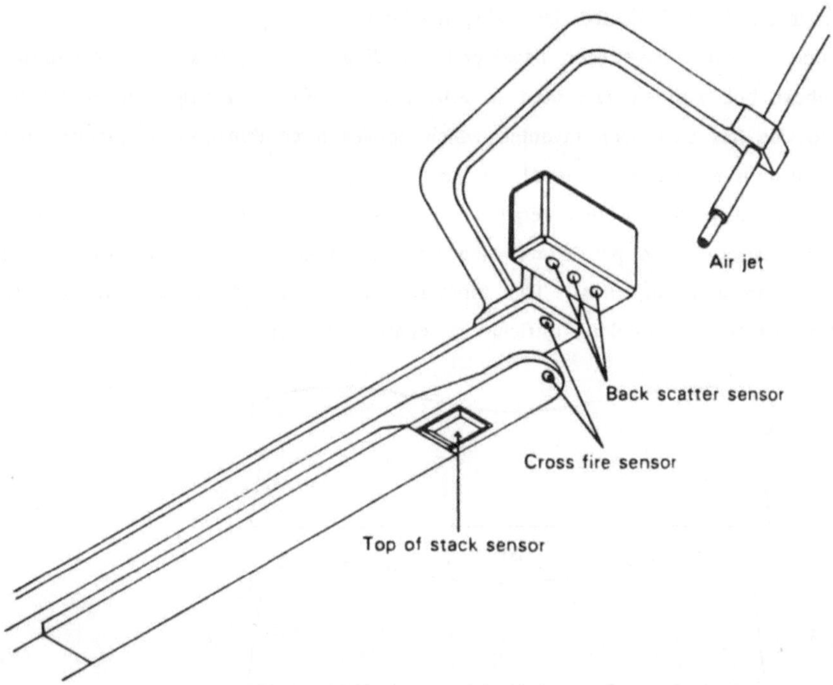

Figure 1 Schematic of an air jet/finger ply separation device.

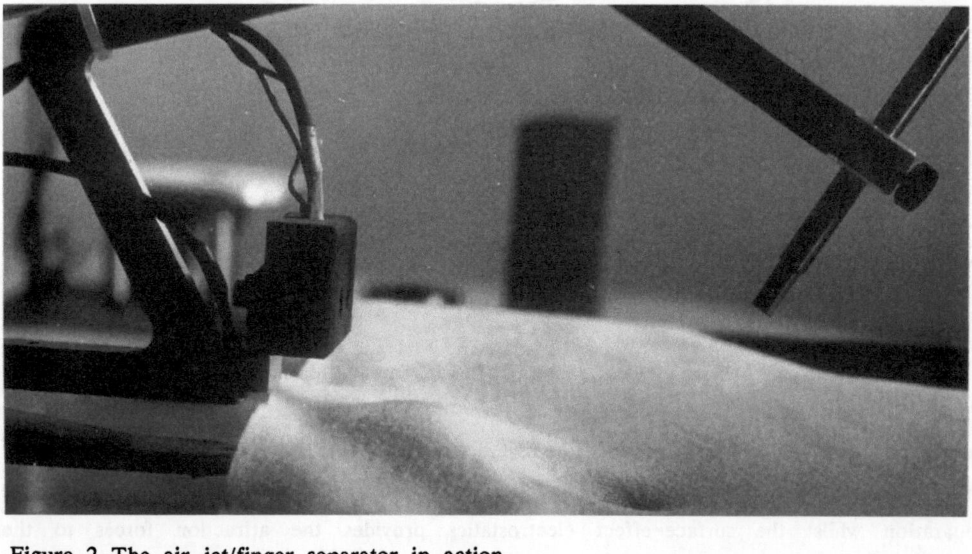

Figure 2 The air jet/finger separator in action.

prime separation means in a dedicated destacking machine.

Other pneumatic device have been developed including an air jet/vacuum mechanism, similar to the above but with vacuum used to attract the fabric to a single upper finger. A vacuum bar device has also been invented which applies a combination of vacuum and mechanical manipulation to the top edge of the fabric to achieve separation.

The shirt collar assembly project requires the destacking of one of each of three collar panels from stacks. These panels are then inspected and placed one on top of the other for feeding into a fusing press. The topmost panel is slit down the middle, see figure 3, and this makes it particularly difficult to separate and handle.

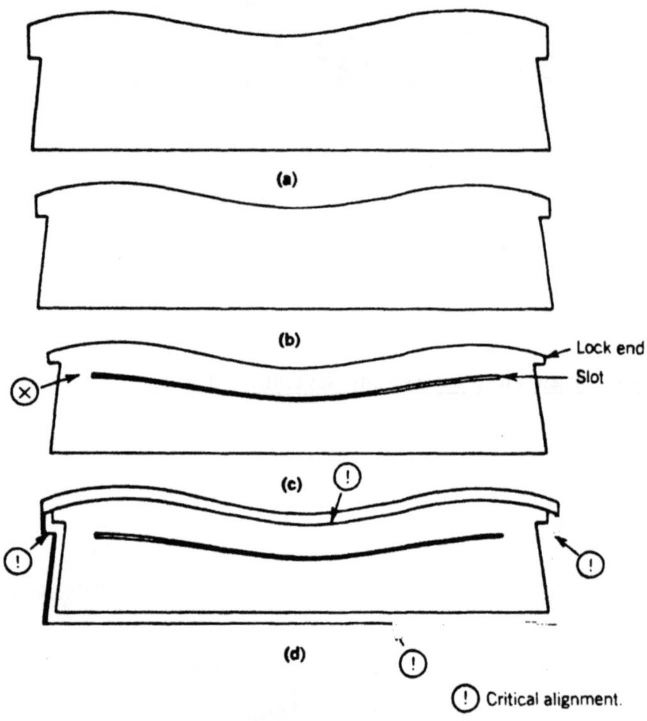

Figure 3 The component parts (a–c) and the completed (d) shirt collar construction.

Electrostatic separation devices have been invented (Monkman and Taylor, 1988, Taylor and Monkman, 1988) which perform the task as required. The gripper shown in action in figure 4, is based on an electrostatic roller. This is rolled over the stack and pressed down whilst a high voltage is applied to its surface. The rolling action aids edge separation whilst the surface–effect electrostatics provides the attraction forces to the topmost ply.

Figure 4 Electroadhesive ply separation.

PICK AND PLACE

The first problem considered was to try to extract an embroidered motif (which could be one of many designs) from a bin and place it in the desired position and orientation on a separated panel of fabric. The approach taken was to perform a blind lift of motif(s) out of the bin, and then place the motif(s) on an inspection table. A vision system then analyses the image to detect firstly if there is a single motif, the correct way up, if there are too many motifs, if there is a single motif upside-down, if there is a motif only partially within the field of view, or if the motif is of the incorrect design. The measures used are area and perimeter. If any of the error conditions occur, the robot is instructed to take appropriate action such as 'try to remove surplus motifs' or pick up the motif and place it in a turning-over box for further presentation. If a single motif, the correct way up is found then it is oriented before pick-up. This is done by rotating the table and repeatedly finding the exclusive OR (pixel difference count) between the actual image (centred in software) and a taught image in the correct orientation (Taylor G.E. et al, 1982). The time constraint was 10 seconds to match the parallel fusing time so a simple slow approach using cheap hardware and simple software proved to be quite effective.

The second pick-and-place task considered forms part of the underwear assembly project. Here, it is necessary to lift one separated item, such as a gusset subassembly, and place it onto the front panel prior to sewing them together. Figure 5 shows the experimental gripper used to test the idea of using multiple low-cost, gripper-mounted cameras (Taylor P.M. and Bowden, 1986). The sequence of operations is illustrated in figures 6 and 7.

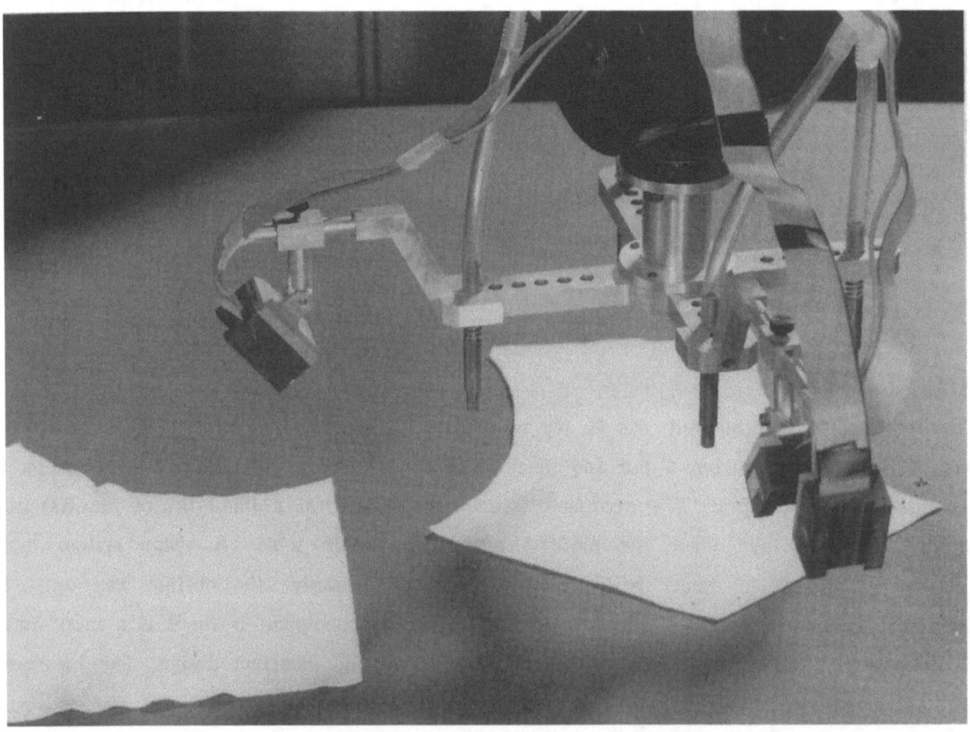

Figure 5 Picking up a gusset panel.

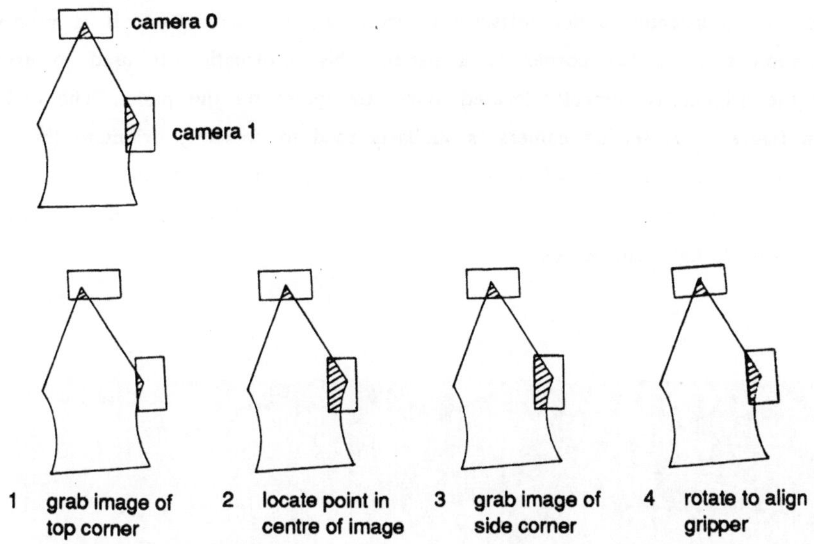

Figure 6 Vision–servoing operations to align the gripper over the gusset panel.

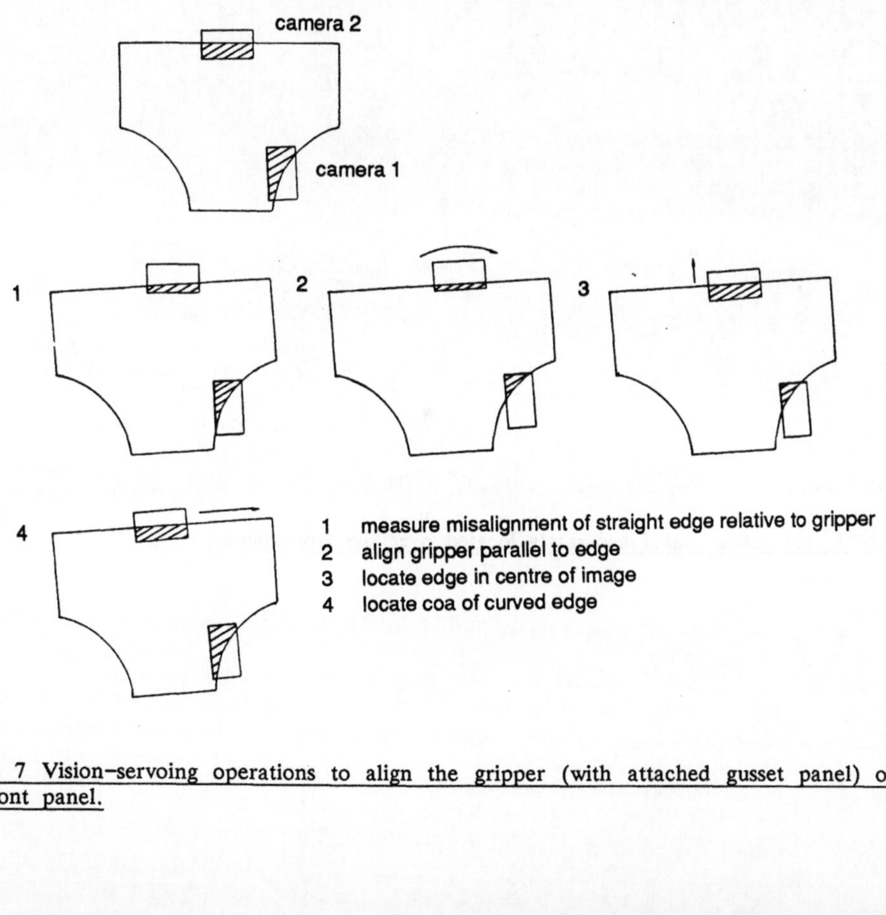

Figure 7 Vision–servoing operations to align the gripper (with attached gusset panel) over the front panel.

A camera is selected under software control, and the data from it is processd to find information such as the corner of a panel. This information is used to drive the robot until the gripper is correctly located above the point on the panel. The end result is shown in figure 8. A second camera is similarly used to correctly orientate the gripper. This procedure is carried out before picking up the gusset subassembly and again just before placing it on the front panel. The cameras are DRAM types which have been rehoused in rugged, light enclosures.

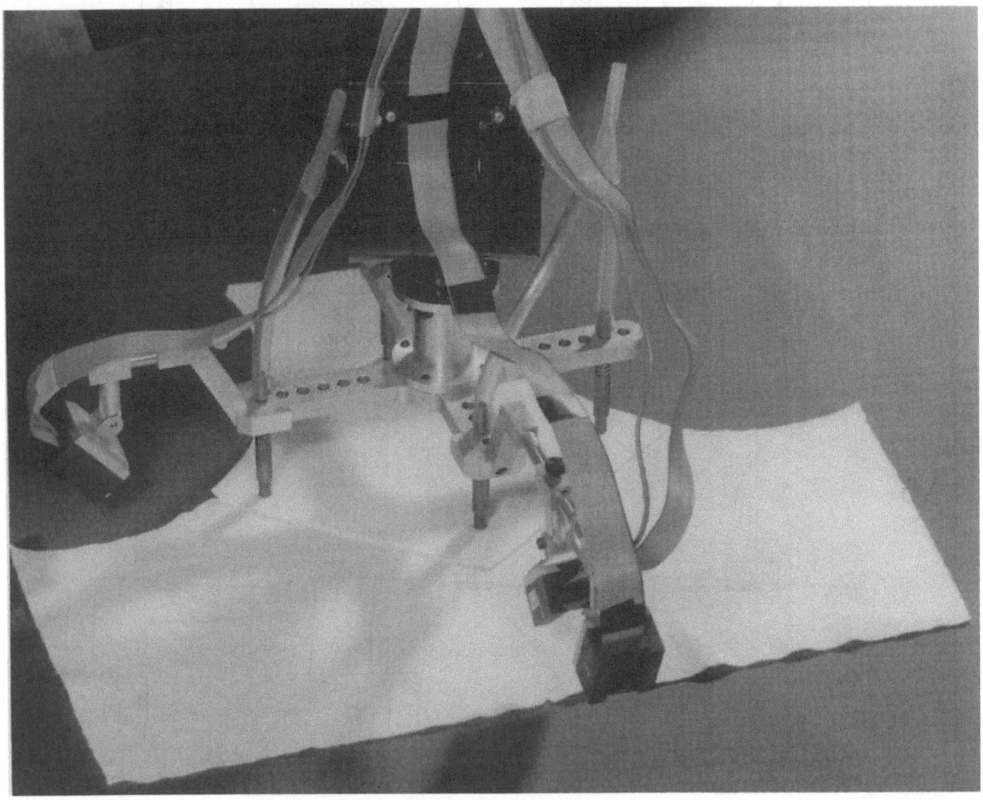

Figure 8 The gusset panel accurately located over the front panel.

2D MANIPULATION

The main 2D manipulation task carried out so far has been the sewing of binding tape to the edges of underwear components. It is necessary to hold the fabric panels securely whilst feeding through the sewing machine. This has been achieved by hollowing out the lower jaw of the air jet/finger ply separation device and applying vacuum. The robot movements are slaved to the reciprocation of the sewing head, although this has meant modifications to the UMI RTX robot's controller to obtain the required accuracy of continuous path motion (Taylor and Gilbert, 1988). A second approach, seen in figure 9, uses a flat plate electroadhesive gripper to hold a fabric panel during sewing. The

Figure 9 Sewing using a robot with an electroadhesive gripper.

limpness of the fabric is a mixed blessing. On one hand it is unnecessary to halt the robot movements whilst the needle is down. On the other hand fabric control can be difficult, particularly at the trailing edge.

3D MANIPULATION

The closing processes result in a three dimensional object which cannot be flattened into a two dimensional object keeping one surface uppermost. If a particular operation, such as putting in a waist band, has to be carried out, then the edges to be handled must be moved locally or globally so that the attachment can be made in two dimensions (locally) at the sewing head. This requires some kind of three dimensional manipulation and is a feature of a current research project.

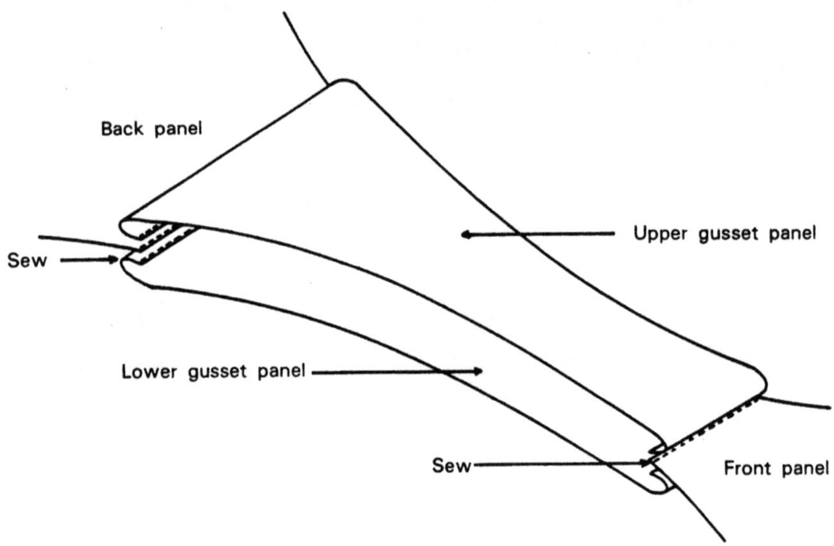

Figure 10 The double concealed gusset construction.

The underwear subassembly project looked at the task of producing a double concealed gusset construction, shown in figure 10. The sequence of operations is shown in figure 11 and are described in more detail in Taylor and Koudis (1988). The approach is to first construct the subassembly inside-out and then turn this inside-out to give the required result. The front and back panels are rolled up on a pair of rollers attached to a robot. Infra-red sensors are used to stop the rotation before the trailing edge of the panels are reached. The rollers are then placed on top of a separated bottom gusset

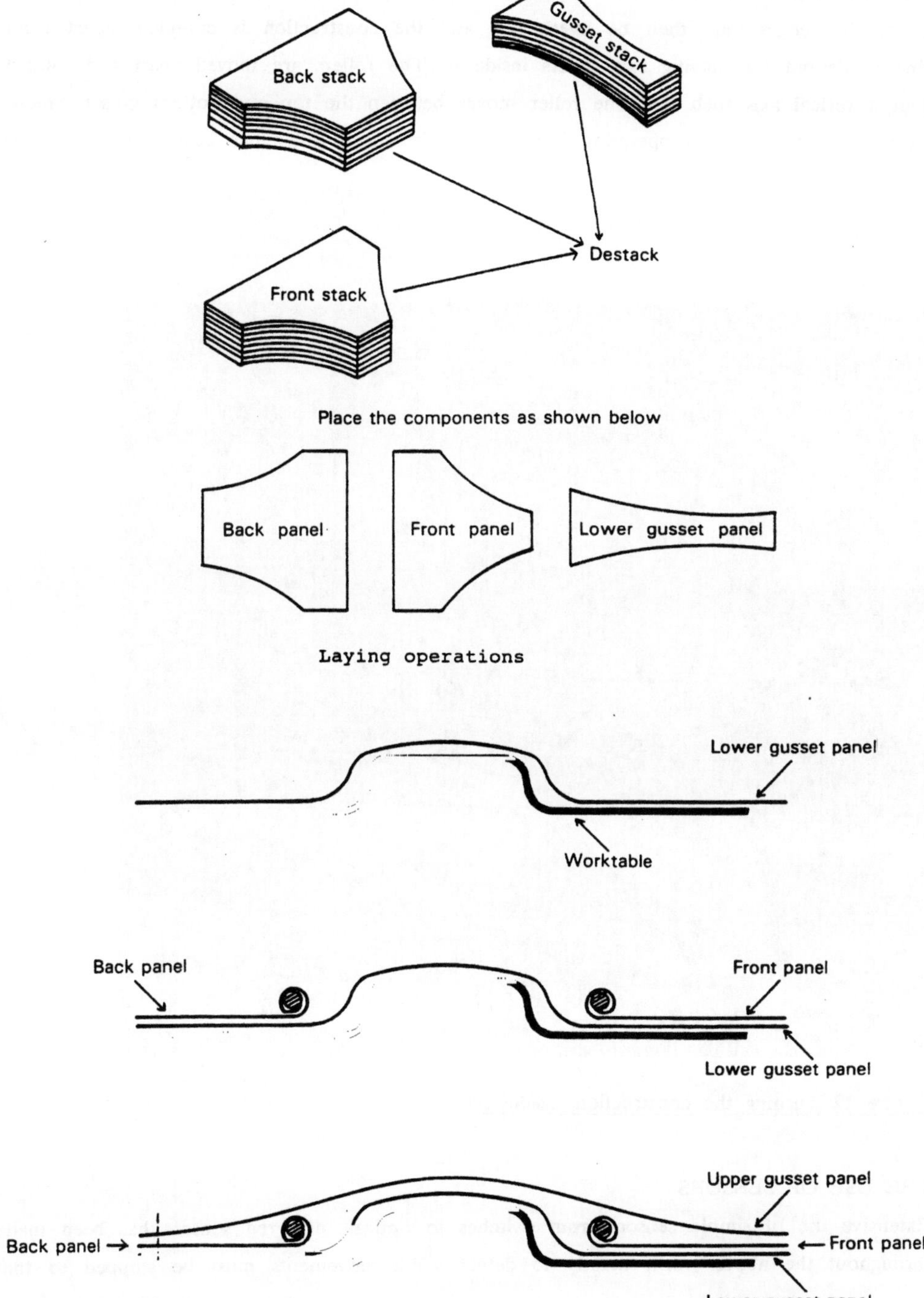

Figure 11 Laying-up operations prior to sewing.

panel. The top gusset panel is then placed on top of the rollers with their front and back panels. The edges may then be overlocked and the construction is complete apart from being inside-out and having two rollers inside it. The rollers are moved apart and rotated about a verical axis such that one roller moves between the top and bottom gusset panels, figure 12. The inside-out operation is complete after each roller has been thus rotated by 180 degrees.

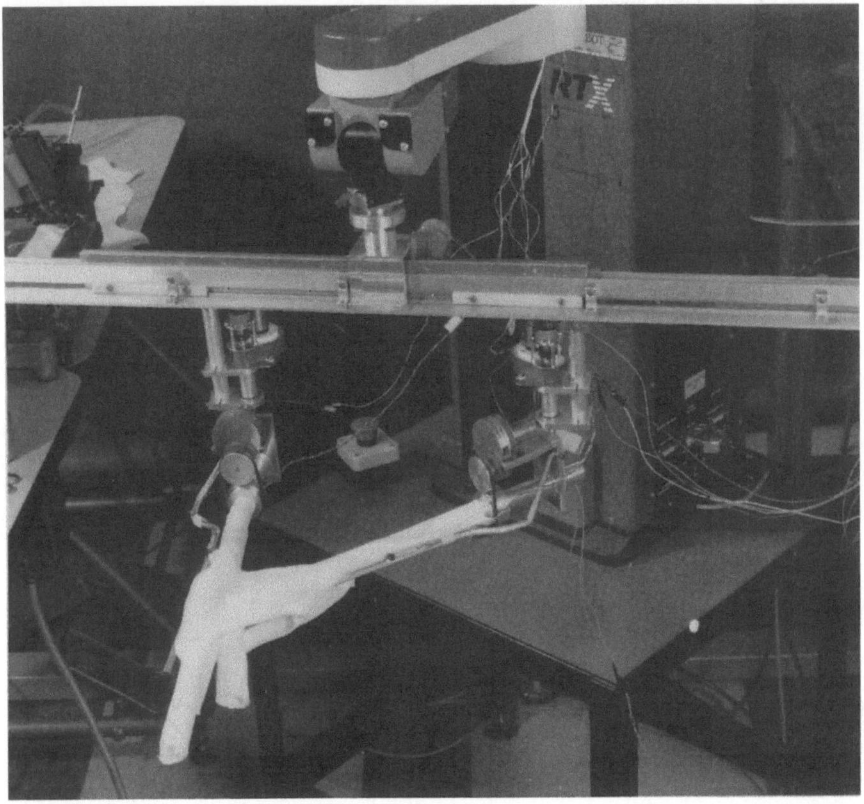

Figure 12 Turning the construction inside-out.

THE USE OF SENSORS

Extensive use of simple sensors from switches to optical infra-red sensors has been made throughout the above work, mainly to detect when movements must be stopped so that edges are in the right place. Two dimensional vision has been implemented in the pick and place operations, again simplicity has been aimed for, with effort put in to structure the problem and the environment so that simple processing techniques can be applied. Sensors have also been used to monitor the various operations being carried out and the information used to select appropriate error recovery routines so that the desired reliability can be achieved. Much needs to be done in this area, particularly around the sewing operation and during future 3D manipulation of fabric. It may well be that to get high reliability, the fabric behaviour will need to be sensed by a mixture of visual and tactile sensors and adaptive control techniques used in the control of the manipulating machinery.

Another area of research lies in how to synthesise complex sensory systems such that a wide variety of garments can be produced with minimum change-over time. A sensory robotics system will be of no use if it takes sevaral man years of PhD calibre manpower to put a cell together which then has to be changed after a morning's work. These difficulties are addressed in a later paper in the workshop.

REFERENCES

Dlaboha I. (July 1981) Cluett's Clupicker enhances robot's ability to handle limp fabric, *Apparel World, Knitting Times.*

Kemp D.R., Taylor G.E., Taylor P.M. and Pugh A. (April 1982) A sensory gripper for handling textiles, *Proc. 13th Int. Symp. on Industrial Robots, Chicago.* This is also available in *Robot Grippers*, eds. Pham D.T. and Heginbotham W.B. (1986), IFS (publications) Ltd, pp. 155-164.

Monkman G.J., Taylor P.M. (April 1988) Electrostatic grippers: principles and practice', *Proc. Int. Symp. on Industrial Robots, Lausanne.*

Murray J.M. (December 1975) Single-ply pick-up devices, *AAMA Apparel Research Journal,. Bobbin.*

Taylor G.E., Kemp D.R., Taylor P.M. (1982) Vision applied to the orientation of embroidered motifs in the textile industry, *Proc. 2nd Int. Conf. on Robot Vision and Sensory Controls, Stuttgart.*

Taylor P.M., Bowden, P. (November 1986) The use of multiple low-cost vision sensors in fabric pick-and-place tasks, *Proc. IFAC Symp. on Low Cost Automation, Valencia*, pp. 89-95.

Taylor P.M., Gilbert J.E. (August 1988) Computer aided design of variable structure controllers for robots with drive nonlinearities, *Proc. 4th IFAC Symp. on Computer Aided design in Control Systems, Beijing*, pp. 145-150.

Taylor P.M., Koudis S.G. (1988) The robotic assembly of garments with concealed seams, *Proc. IEEE Conf. on Robotics and Automation, Philadelphia*, pp. 1836-1838.

Taylor P.M., Monkman G.J., Taylor G.E. (1988) Electrostatic grippers for fabric handling, *Proc. IEEE Conf. on Robotics and Automation, Philadelphia*, pp. 431-433.

Projects for the Garment Industry at the Fraunhofer Institute for Manufacturing Engineering and Automation (IPA), Stuttgart

Holger Nollek
IPA
Nobelstr. 12
7000 Stuttgart 80

Introduction

Since the economic situation is forcing the clothing industry to consider the possibilities of improving competitiveness, an increasing number of companies are no longer excluding automation from their production out of hand. In order to accelerate the automation, which is planned and which is certainly also necessary, it is absolutely essential that co-ordinated efforts are made to guarantee the efficiency of the equipment used and the development of practical automation components. Assistance in answering the question of what is to be developed and when, may be provided on the basis of experience in other branches of industry which have already undergone similar development. A local determination of clothing production is carried out in the following, then conclusions drawn, from a comparison with developments in other branches of industry, with regard to sensible research and development activities, current and future projects at the Fraunhofer Institute for Manufacturing Engineering and Automation are explained, then a forecast of further development provided.

Local determination

Let us consider clothing production as it would be if the most up to date technology available were to be used. For this purpose we may logically divide production into the four production areas: storage, cutting, sewing, and finish (see Fig. 1). The processing machine in the cutting department is the computer controlled automatic cutting machine. The handling

NATO ASI Series, Vol. F 64
Sensory Robotics for the Handling of Limp Materials
Edited by P. M. Taylor

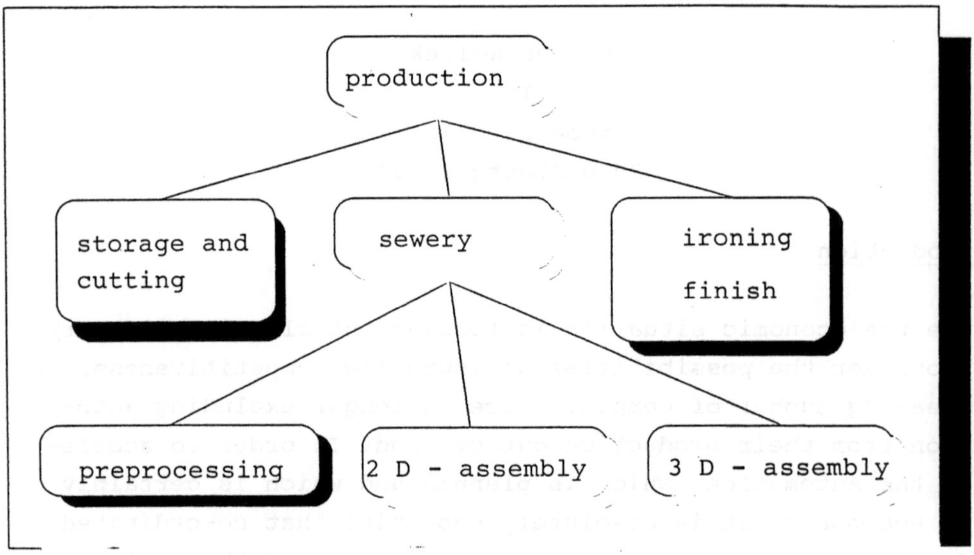

Figure 1: Structure of garment production

operations to be carried out before and after machining, i.e. after cutting, include the laying of the layers to be cut, and the clearing of the cut stacks and discharge of waste. Whilst solutions which are automated to the greatest possible extent are already available for the laying of layers the clearing of cut stacks is still a manual operation.

In the sewing area development in recent years has concentrated mainly on the sewing machine. This has resulted in sewing speeds which make it almost impossible for the seamstress to guide the sewn material during machining. As the next logical step came the development of automatic sewing machines, which now only require manual tightening of the workpiece, otherwise their operation is completely automated.

In recent years the transport systems, some of which are already provided with trolley identification, destination control etc., and may therefore be considered components of automated production, have also undergone considerable improvement.

Because of the complexity of the problems to be solved in automation, the fields of ironing, finishing, and final storage should not be considered as long as production cannot be satisfactorily automated in the areas in which the workpiece is still 2-dimensional.

Conclusions

If developments in other branches are now compared with the situation in clothing manufacture it will be established that automation of the handling processes in the area surrounding the sewing machine has largely been omitted for reasons which are very well known. However, it is in just such areas that the biggest economic successes may be expected, since up to 70 % of all production time is spent in handling operations.

What does this mean for future development?

According to the above division of production into storage, cutting, sewing, and finish, it is certainly reasonable to develop systems - in the cutting area - which load the cut stacks into suitable transport containers.

This would create the conditions for automatic processing of the stack in the sewing area. Sewing as a part of the whole production (Fig. 1) can be further divided into pre-production, in which components are prepared for subsequent assembly, the assembly of flat parts, and the assembly of 3-dimensional parts. Due to the problem in handling 3-dimensional bodies of pliable material, this field should not be included in the considerations until the problem of handling flat parts has been satisfactorily solved.

The workplaces in the pre-production and 2-D assembly may be characterised on the basis of the condition of the prepared workpiece stacks:

- cut stack only
- at least one stack already isolated.

The resulting requirements regarding the automation components are striking: because of the exact contour of the stack, automatic separation and positioning in front of or under the needle, is quite feasible with the systems currently available. The situation is more difficult, however, if the workpiece stack has already been separated during previous machining, and has been built up again by a stacker. The position inaccuracies of the workpieces in such a stack necessitate a unit for positioning and orientating the parts in front of the needle.

When two parts are assembled the operation of bringing together the workpieces to be sewn is added to this. If the structure of these workplaces is considered, the sub-systems of an automatic sewing cell as shown in Fig. 2 are obtained. Magazines for receiving the workpiece stacks are in use in a wide variety of forms in other branches of industry. Systems for positioning, orientating and placing workpieces one on top of the other are

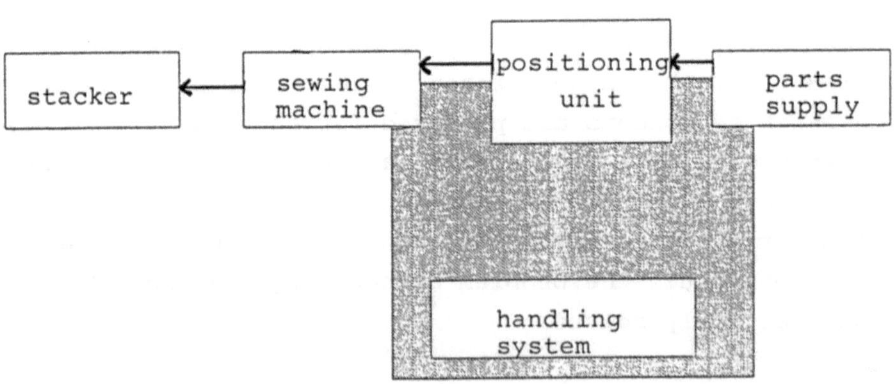

Figure 2: Components of an automated sewing cell

already on the market or are being developed, and will soon be available. On the handling systems it must be said that they represent the state of the art, and that on the German market alone well over 200 different units are offered, with a wide variety of different characteristics and features. The systems suitable for the clothing industry can certainly be selected from this wide range of supply. The gripper is an inseparable part of the handling system. Even if there will probably never be a gripper which is able to pick up all types of materials with the same degree of reliability, a suitable gripping system can certainly be found for almost every material. The required gripper can automatically be changed by means of gripper changing devices, which are at present state of the art. The sewing machine is also a well known component, although it is now undergoing one or two modifications. Thus it is essential that the central computer is provided with information about the system conditions in order that spool, thread and needle changes can be carried out automatically. One sub-system which requires a complete overhaul is that of the stackers. the stacker of the future will no longer throw the parts in a disorderly manner through a bar or into a box, but will store them in an orderly fashion in a magazine. Only thus is it possible to continue the process in the following sewing cell without the insertion of a manual sorting position.

If it is assumed that automated sewing workplaces should not represent isolated islands in production, then the question of the linkage between the workplaces remains to be answered. The above-mentioned magazines, which are filled by stackers, then conveyed to the next sewing cell, certainly provide the possibility of this (Fig. 3).

The development of information processing will run parallel with the development of these components, which will take about 5 - 10 years until they can be regarded as the state of the art in the clothing industry.

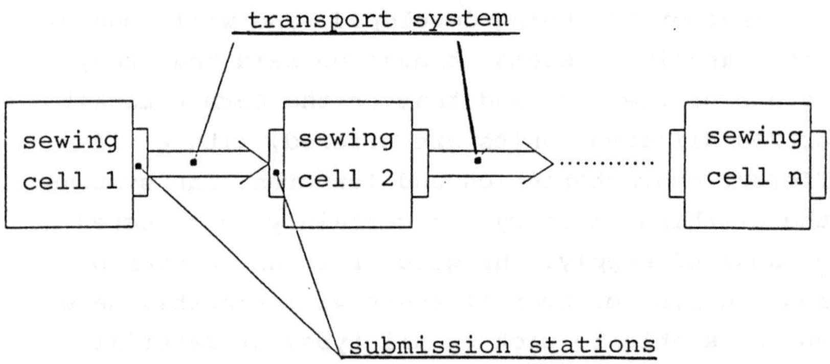

Figure 3: Material flow in the sewing room

Additional information of importance to the production process can be obtained using CAD systems for cutting pattern design. Thus the information important for the sewing machine, on which thread colour should be used and when, can be obtained - on the basis of the magazine number - from the data determined during drawing up the cutting pattern in the CAD system. This also applies to the information on the stack structure, i. e. in what sequence which types of materials of what thickness are to be processed. On the basis of these data the gripper can be automatically changed by the handling system and the needle penetration depth can, for example, be adapted to the material thickness.

Summing up it may be established that the following components must be developed:

- Systems that separate the material which can be used from waste material after cutting
- Magazines
- Improved stackers for filling these magazines
- Material flow system for linking the workpieces together
- Automatic sewing machines with automated secondary functions
- Sewing machines with replacable sewing heads
- Positioning devices for preparing the workpieces before they reach the sewing machine.

Essentially, this list also represents the research projects of the IPA in Stuttgart which are described in the following.

Projects at the Fraunhofer Institute for Manufacturing Engineering and Automation, Stuttgart:

In the field of clothing production, there are currently 4 projects worth mentioning, one of whch is nearing completion. The other three will commence in the near future.

CAW - Computer Aided Waste Removal

In this project a system will be developed for separating the cut parts from the waste, and for supplying them in a defined position, in a magazine, on a conveyor belt or similar. Here it does not matter whether individual layers or stacks are cut.

A precondition for the installation of such a system is the presence of a CAD system and an NC cutting table. The structure of the entire system is shown in Fig. 4. The movement program of the NC cutter - generated by the CAD system - is supplemented by additional cutting edges in the CAW module in order to have an adequate manner of dividing up the waste. Furthermore, the parameters required for controlling the handling system are obtained, the movement program generated and then transferred to the control system for the clearing unit.

In addition to the software for the CAW module the kinematics of the handling system, the required grippers for individual layer and stack access as well as the interfaces to the transport system must be developed within this project.

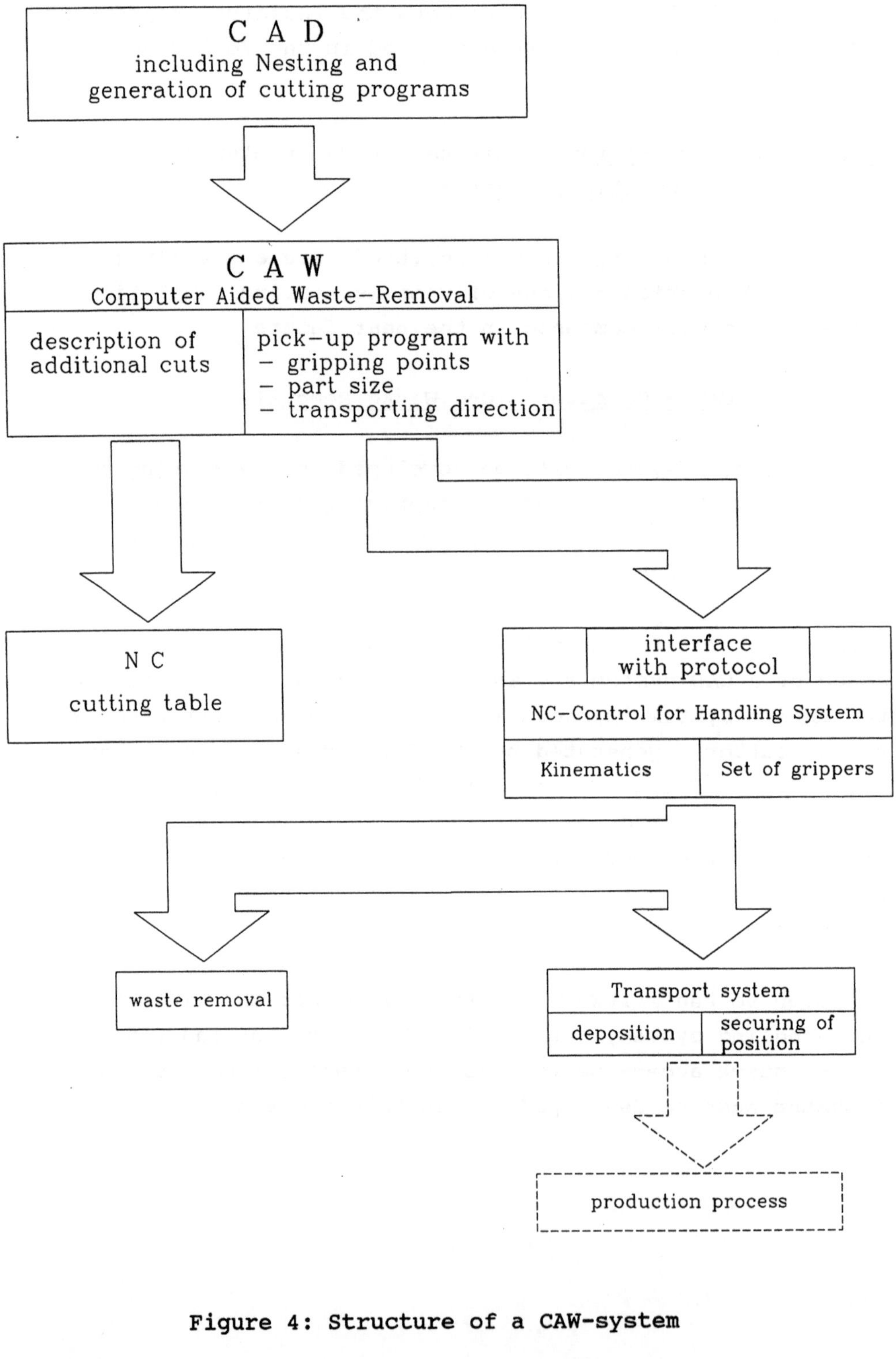

Figure 4: Structure of a CAW-system

Sewing cell

A flexible sewing cell has been developed for work pieces in collaboration with the Hohenstein Clothing Physiological Institute, Bonnigheim.

The entire system consists of the following components (see Fig. 5):

- Magazine unit for two magazines
- Positioning system for flat workpieces
- Two-axis handling system
- Commercially available sewing machine

The prototype of this sewing cell is designed for the production of trousers. Depending on the type of sewing machine used, either single parts can be stitched or the side seams of trouser legs can be closed. The positioning unit, the centrepiece of the system, is also able to position and, if necessary, without manual adaptation, place workpieces of different sizes and materials one on top of the other. This provides the possibility of trouser legs, sleeves, coats, jacket parts and much more to be processed. The positioning accuracy is approx. ± 1 mm, and therefore meets the reqirements of the clothing industry.

The sewing cell operates on the basis of a user-convenient man-machine interface, which is provided by means of an IBM PC. Information on the condition of the system is constantly displayed on the screen. For example the current machining step, the next machining step, fault conditions and instructions for their elimination are displayed to the operator, together with general information such as the number of parts machined, etc.

As long as it is not possible to obtain parameters such as material type, material thickness etc. from a production host computer, the information is entered by the operator via the PC.

The sewing machine in isolation does not meet the requirments which must be imposed on the components of an automatic production cell. The IPA is therefore engaged in another project dealing with

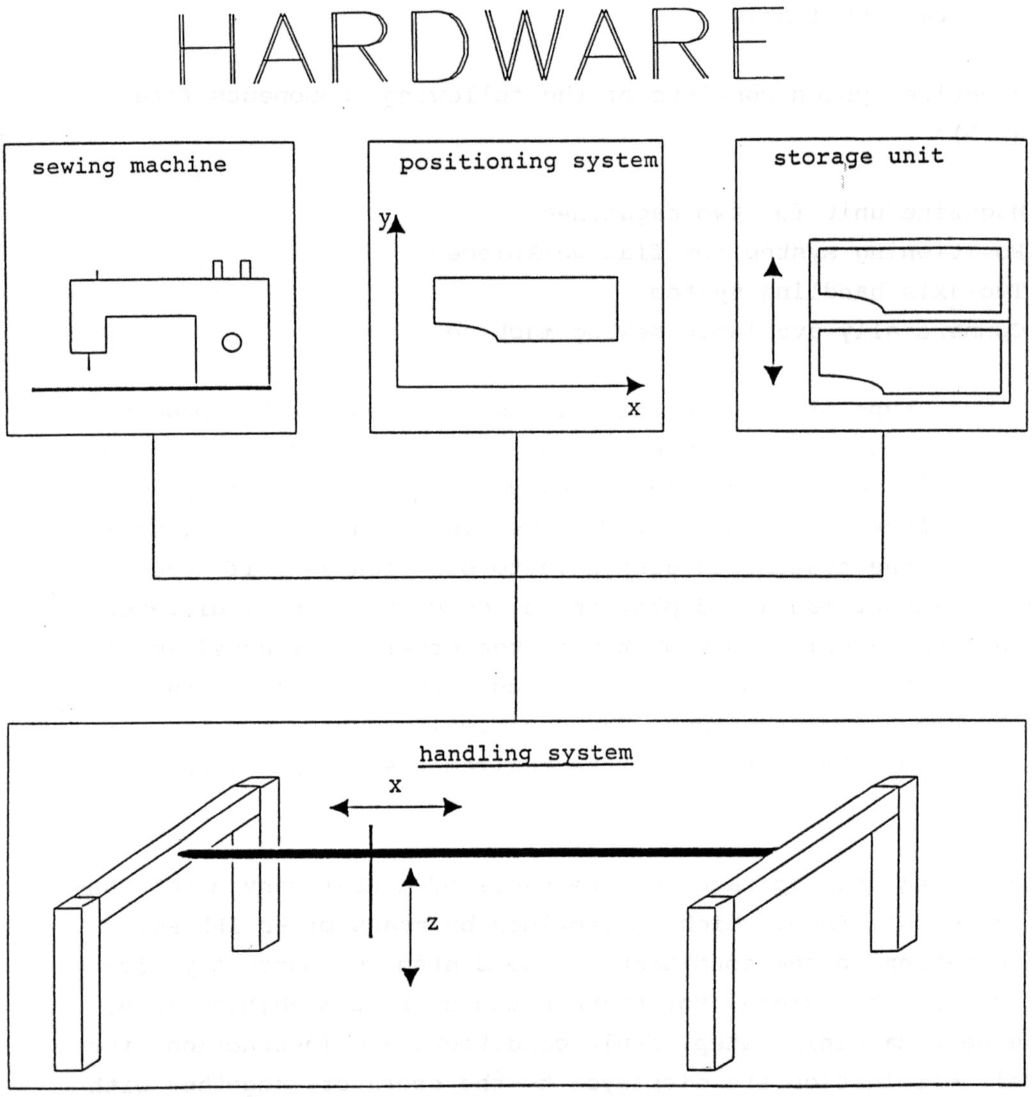

Figure 5: Components of an automated sewing cell

Automation of secondary sewing machine functions

In order to ensure that the sewing machine, the central means of production, is able to meet the requirements of an automatic sewing cell, it must be achieved that those of its functions which are still undertaken by the seamstress, can be carried out automatically.

These functions include, in particular:

o change of the thread in the case of
- color change
- thread breakage

o change of the needle in the case of
- needle breakage

o external inputs for
- Start/Stop
- Reset
- Thread tension
- Speed, upper/lower conveyor
- Spool change
- etc.

o Supply of signals on
- Thread tension
- Thread colour
- Thread supply
- Set speeds for upper/ lower conveyor
- etc.

The sewing machine cannot be integrated in a CIM concept until these functions can be automated.

This project will be implemented together with sewing machine manufacturers and users with the aim: to develop a prototype of a CIM-capable sewing machine within two years.

Cloth Inspection

In recent years the IPA has gained some experience in the application of image analysis and recognition methods in fields which are closely related to the garment industry.

Several projects have been carried out or are current in the image analysis group which is concerned with:

- Failure detection and texture classification of wood mosaics (Fig. 6) - (current work)
- Classification of irregular fibres; in particular: geometric classification of tobacco fibres by means of autocorrelation functions (Fig. 7, 8) - (measurement device developed for a German factory; patent submitted)
- Geometric classification of irregular spray pictures by fractal measures (Fig. 9, 10, 11) - (BMFT project, terminated)

All these activities are based on methods which can easily be transferred to other problems where structures in surfaces need to be automatically inspected.

Beside these practical experiences the IPA image analysis group also has theoretical experiences in most standard as well as modern texture analysis methods. These methods mostly are available from the software package of the group (s. e. g. /1/). The package is written in higher languages. For test reasons the algorithms can easily be implemented in standard vision development systems.

In the near future we will set up a project for the aparel industry, where we will use the described image and texture analysis methods for automatic cloth inspection.

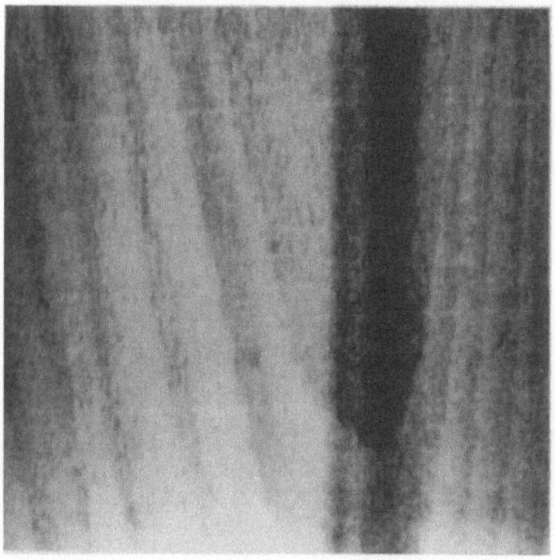

Figure 6: Wood with special defect

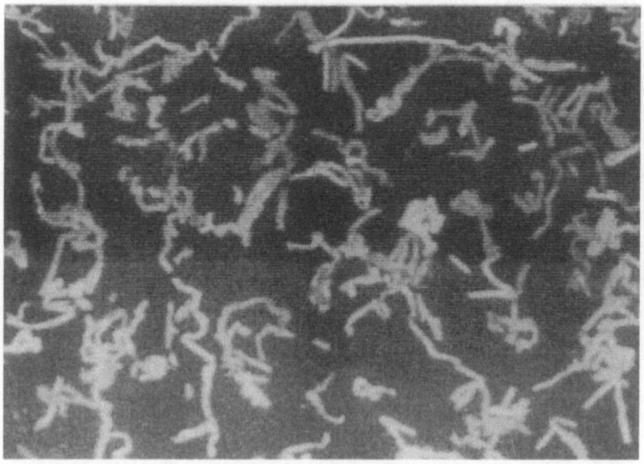

Figure 7: Tobacco fibres grey image

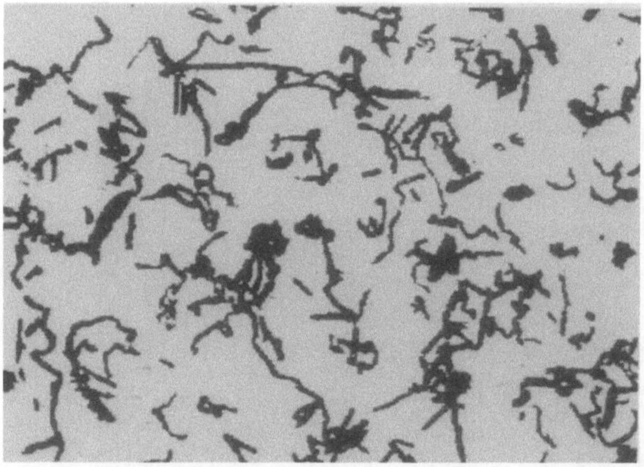

Figure 8: Tobacco fibres binary image

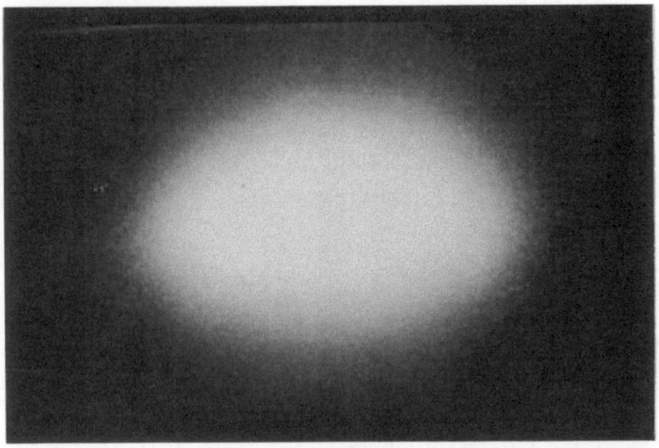

Figure 9: Spray picture - original grey image

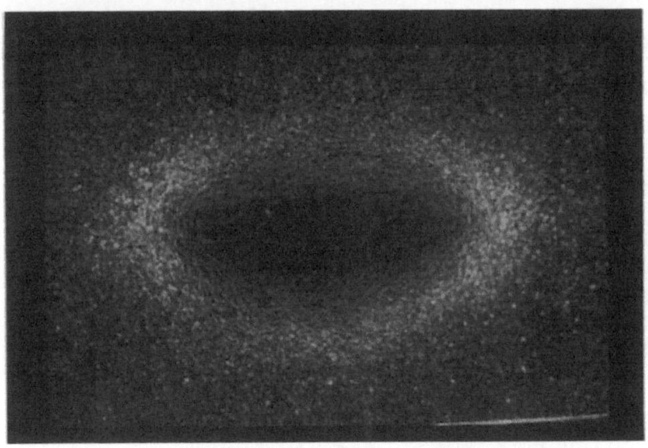

Figure 10: first step of separation of contour and kind

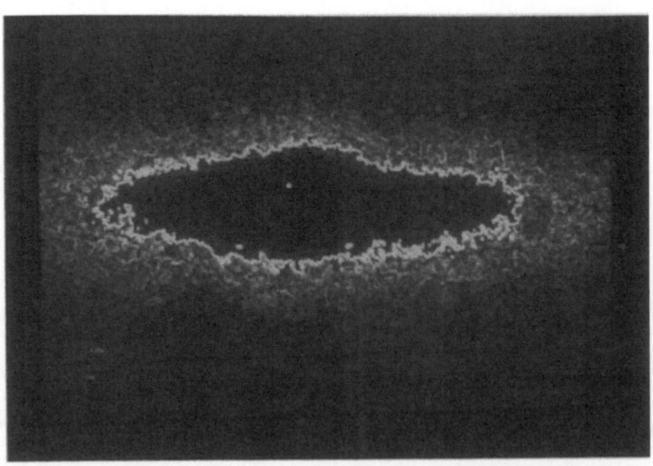

Figure 11: second step of separation

Prospects for future development

The future of the clothing industry, in the so-called high wage countries, depends essentially on the success of the rationalisation efforts being made. National and international research activities clearly indicate the importance of this subject. The Fraunhofer Institute for Manufacturing Engineering and Automation, Stuttgart, will also continue its efforts in the development of automation components for the clothing industry. After the current and planned projects, all of which are aimed at developing hardware components, the data integration of all the production components into one CIM concept will be the central objective.

When the sewing cell is being fed with full magazines the cell computer can call up information about the content from the host computer on the basis of the magazine number. The corresponding grippers, thread colours, thread tension, etc. can be prepared inside the sewing cell based on the information obtained about type of workpiece, part dimensions (e. g. waist-band width, trouser parts), sequence of material types and colours in the stack, etc.

These data were obtained when the programs for automatic layer and cutting machines were being generated. After machining the cell computer signals the end of the machining stage to the host computer, thereby providing a continuous documentation of the progress of the machining.

ANALYSIS OF A PLY-SEPARATION GRIPPER

F. VERCRAENE P. ESQUIROL
Ecole Nationale Superieure des Techniques
Industrielles et des Mines d'Ales
30107 ALES FRANCE

ABSTRACT

Analysis of textile prehension: an attempt is made to decouple the problems associated with textile prehension using, as an example, general garment production, giving examples of existing grippers.

INTRODUCTION

The problem of textile process automation is enormous, and garment factories are frequently automated by an internal department of a company, often resulting in inflexible automation solving a local problem only. Indeed, many systems of destacking, transportation and stacking have been realised in garment factories having the incontestable advantage of being inexpensive, however they cannot be transposed from one product to another because they take account of the shape, size, nature and sometimes the colour of the material. We must take a more general step to increase the flexibility of the production methods and also attend to the garment path in the assembly line.

We can, in general, consider the assembly line to consist of four successive functions:

- manipulation of pieces in 2D.
- assembly of pieces in 2D to realize an element in 2D or 3D.
- manupulation of pieces in 3D.
- assembly of pieces in 3D to realize an element in 2D or 3D.

NATO ASI Series, Vol. F 64
Sensory Robotics for the Handling of Limp Materials
Edited by P. M. Taylor

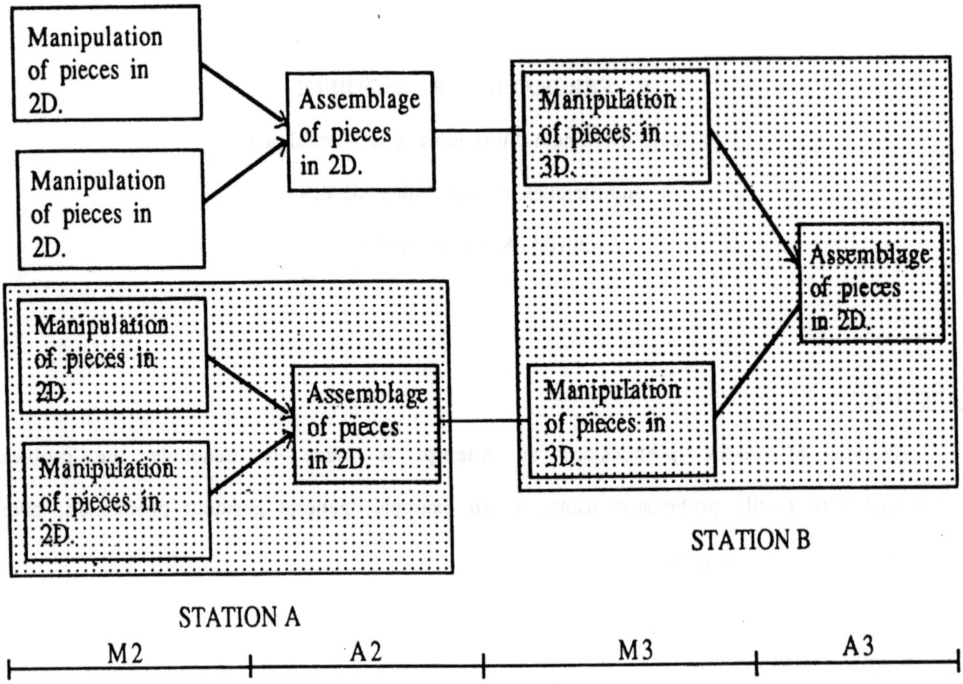

Figure 1:THE FOUR BASIC FUNCTIONS IN AN ASSEMBLY LINE.

The analysis of workstations allows us to consider the manipulation of fabric as the principle obstacle to automation. We shall limit our study to the manipulation of flat pieces, which represent at least 10% of manipulated elements, and up to 100% for some products.

I MANIPULATION IN TWO DIMENSIONS

For a general manipulation task, involving two stacks of pieces of textile, we need the following operations:

- organization
- recognition
- separation
- gripping
- turning
- moving
- positioning
- superposing

Depending on the manipulation device, not all these operations are always necessary. The operations must be decomposed to separate out the problems associated with each.

The 'separation' and 'gripping' operations are known traditionally as 'prehension'. However, when we are speaking about textile prehension we discern four separate actions:

- local prehension
- local separation
- total prehension
- total separation

'Local action' is carried out on the edge of a ply, while 'total action' is carried out on the whole surface of a ply.

These distinctions allow us to break down the problem of prehension, and to pose relevant questions. Moreover, the action of total prehension or total separation appears to be closely related to the problem of ply transport, and in particular to the problem of the control of ply shape.

II LOCAL PREHENSION AND SEPARATION

A preliminary bibliographical study has allowed us to classify the existing systems into wide ranges by comparing the order of local prehension and local separation.

Our classification has as its goal the decoupling of separation from prehension order to simplify the problem. Considering the prehension and the separation we can:

- begin with the separation and then assume the prehension; this is the direct type.
- assume simultaneous prehension and separation; this is the simultaneous type.
- begin with the prehension and then assume the separation; this is the reverse type.

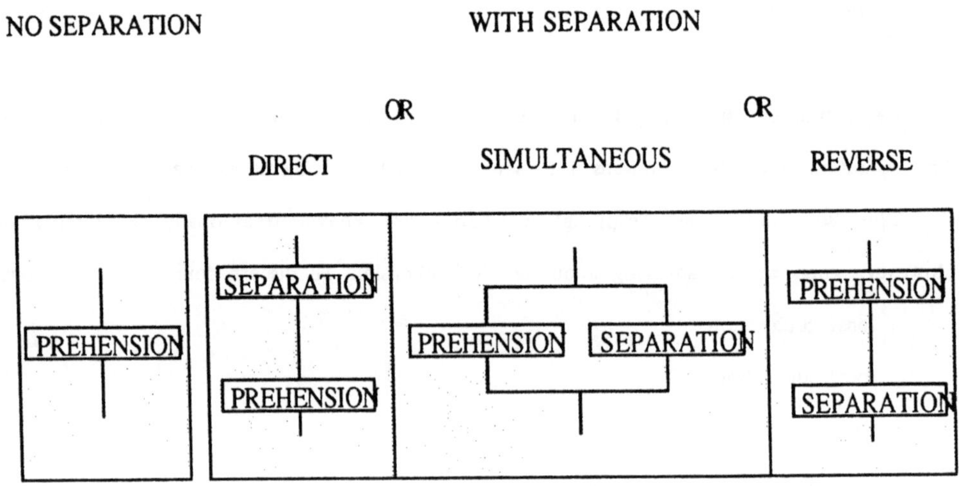

Figure 2:Classification of prehension and separation

III REVERSE TYPE PROCEDURES

The common characteristic of all these procedures is to stick superficially to the first ply of fabric (gripping operation), and then to lift one edge or corner of the ply from the stack, without considering the number of destacked pieces of fabric. Then, another device carries out the separation to ensure a single ply is obtained. After gripping the ply it is necessary to transfer and position it, ensuring that it is flat. The grippers work in a cyclical process composed of the following steps[1]:

...grasp ... transport ... position ... deposition ...

... return to the beginning of the cycle...

To obtain this cyclical process we have to hold the stack to prevent its destruction. After the separation the discarded plies must be replaced on the stack. The following functions illustrate the grasping of a single ply:

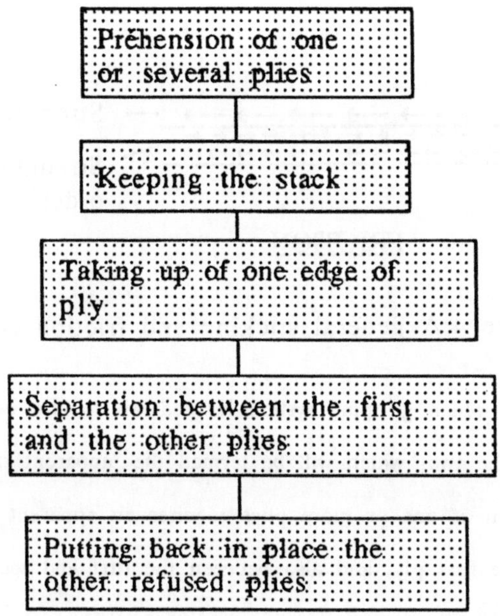

Figure 3:The five operations in a reverse system

<u>PREHENSION OF ONE OR SEVERAL PLIES</u>

In order to grip the fabric superficially we have a choice between a normal surface force, tangential force or a combination of both. For the normal force the sticking coefficient is of importance; for the tangential force it is the friction coefficient which is important.

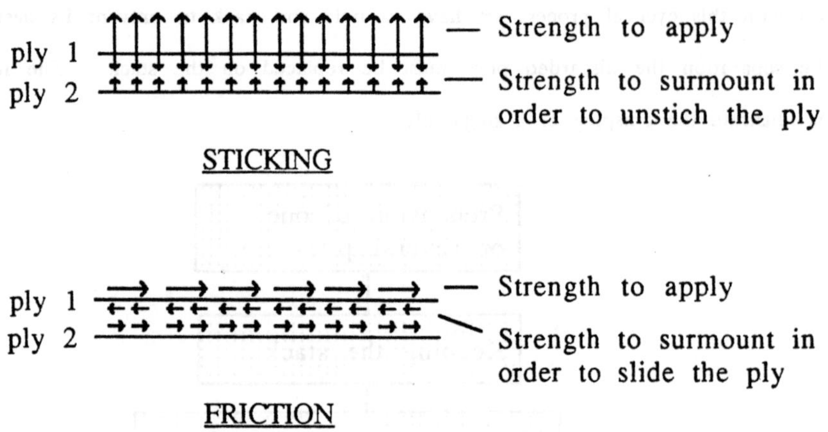

Figure 4:A general sticking and friction coefficient

Depending on which gripping system is used we require one of these coefficients, or a combination of the two, which will be termed the 'apparent sticking coefficient'.

The prehension of one or more plies requires an apparent sticking coefficient between the gripper and the first ply, and also between the first and second ply.

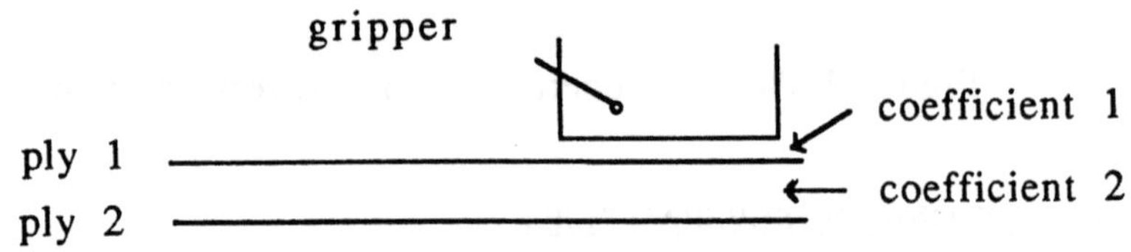

Figure 5:The prehension of one or several plies

The larger the latter coefficient the more reliable the system, however, it is not possible to modify the state of the surface of the fabric. If the gripper modifies coefficient 2 the following inequality must be maintained:

coefficient 1 >> coefficient 2

It is necessary to assume the prehension is carried out without introducing a mechanical device penetrating the plies. The possible solutions are therefore: 'Velcro', vacuum, superficial frost, electrostatic attraction, glass paper for extensible materials, ...

STACK MAINTAINANCE

In order to keep the stack in place, a mechanical pressure, for example, is very reliable and is non interactive with the prehension system.

LIFTING OF ONE EDGE OF THE PLY

For this we must lift one or more plies from the stack. For stitch materials it is possible to achieve this by stretching, unlike other materials.

We will try to lift the ply from the stack with a local displacement of the gripper in order to control the fabric geometry. If we lift the ply with any movement, the ply between the restraining device and the gripper will have an uncontrolled geometry.

SEPARATION OF PLY 1 FROM PLY 2

In order to separate ply 1 from ply 2 we shall use a second gripper that returns (in several operations) the excess plies taken up by the first gripper, leaving only the useful ply.

Then we shall have :

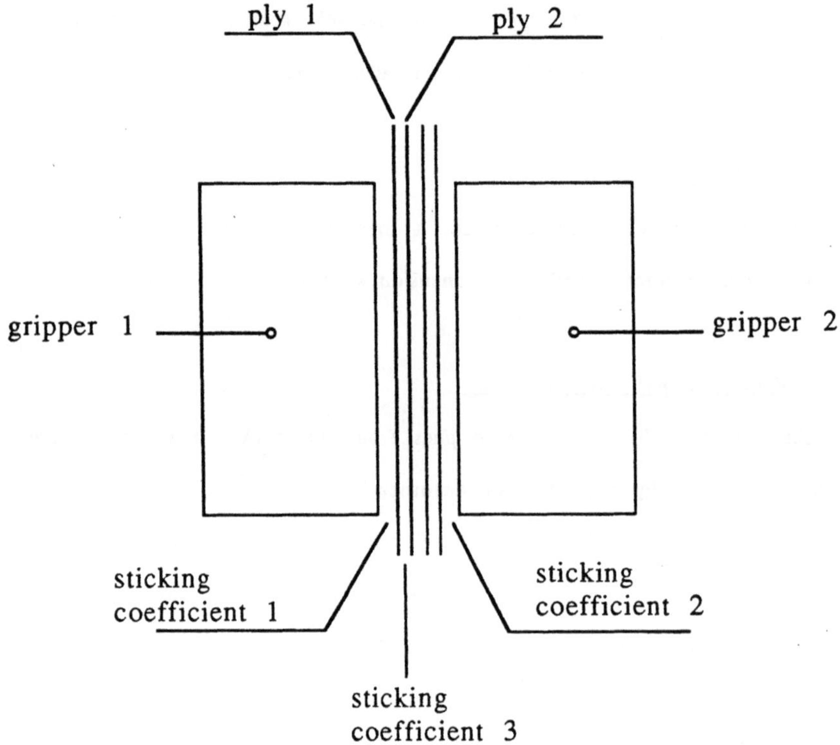

Figure 6: The separation between ply 1 and ply 2

The apparent sticking coefficients have to satisfy the following inequality:

coefficient 3 < coefficient 2 < coefficient 1

For this second gripper we have the same possibilities as for the first. However, the system will be all the more reliable as the difference between these coefficients will be large.

REPLACING THE EXCESS PLIES

This operation is generally achieved using the rigidity of the material and its weight. It is possible to increase this effect using suction beneath the stack.

IV TOTAL PREHENSION - SEPARATION

After the local prehension–separation, it only remains to propagate the separation. Meanwhile, the distortion of the material during the total prehension has to be minimized because it is conditioning, in part, the quality of the final product.

The total prehension–separation is usually effected either by repetition of the local gripper operation, or by single separation of the plies.

TOTAL PREHENSION

In order to analyze this problem we have classified the process of prehension with distribution of the prehension components over the surface of the fabric. The criteria for classification will then be the density of gripping devices on the gripper. This classification (see the following table) is more or less arbitrary because there is a quasi continuous advance of the principles of total prehension. This table does not take into account the general shape of the ply on or under the gripper. In particular, if the fabric is rolled over the gripper it is better able to keep hold of it.

V CONCLUSION

This analysis is the first step in researching the problem of textile manipulation. For system reliability the inequality

$$\text{coefficient 3} < \text{coefficient 2} < \text{coefficient 1}$$

has been widely verified, so that the differences between coefficients are important.

Coefficient 3 (apparent sticking coefficient between plies) is bound to fabric properties; it is stable data that is not kept down. On the other hand, coefficients 1 and 2 are, in the main, process dependant and can be kept down. The best process will be one that allows us to best satisfy the previous double inequality. It is necessary to ensure that a strong ply–gripper adherence (coeff 1 and 2) does not increase coefficient 3 (for example suction) and thus no longer satisfies the inequality.

In order to obtain high reliability and optimal flexibility the double inequality must always be satisfied, whatever kind of fabric is used, and whatever the apparent sticking coefficient between plies.

REFERENCES

[1] L'automatisation en confection. Confection 2000 N° 72 Octobre 1986. F.VERCRAENE.

[2] Installation automatique de prehension de pieces, notamment de pieces textiles. Brevet, France. N° 2349423 1976.

[3] Robot prehenseur Confection 2000 N° 69 Novembre 85.

[4] These: Contribution ã la realisation d'un manipulateur pour materiaux textiles. R. BOUHYA 1986.

[5] Automation of handling and production for apparel industry. Dr. M. SARHADI and Professor M.J.H. STERLING. School of engineering and applied sceince, University of Durham.

[6] Real time robotic control for apparel manufacturing. E. BERNARDON and T.S. KONDOLEON. Charles Stark draper laboratory. Robots 9 Conference proceedings, June 2-6 1985. Detroit, Michigan.

[7] Dispositif de prehension de pieces d'etoffe ou similaires. Brevet, France. N° 7301399 Janvier, 1973.

[8] A sensory gripper for handling textiles. D.R. KEMP, G.E. TAYLOR, P.M. TAYLOR and A. PUGH. University of Hull. 13th International Symposium on industrial robots, 1983, Congres, Chicago.

[9] The third vetimat. Clothing machinery show, Paris. M. Disher. MACL, Juin 84, vol 65, N° 6.

[10] Advances in gripper technology for apparel manufacturing M. SARHADI, P.R. NICHOLSON, and J.E. SIMMONS. School of Engineering and Applied Science, University of Durham.

[11] L'automatisation en confection d'articles maille. J. MESNY. Confection 2000 N° 48 Juillet 1984.

[12] Organes et systemes de prehension. M. EDEL. Point on robotique, Vol. 1. Ed. Lavoisier Tec & Doc.

Round Table Discussion
Sensory Robotics in Garment Manufacture: Trends and Future Needs

Chairman: Prof. F. Paul
Secretary Mr. D.G. Caldwell

The round-table chairman started by saying that based on observations of today's wide-ranging discussion a number of issues seemed to have arisen. He outlined these as;

Mass vs Flexible Production
Product vs Machine Design
Art vs Engineering Approach
Textile vs Apparel
2D vs 3D Handling

and asked "Is there anything else which should be included?".

It was suggested that research vs development should be included, and there was general agreement that this could most easily be divided along the lines of university versus industry, with research being the domain of the universities and industry being responsible for product development. There did, however appear to be a problem in defining where the university role ended and the industrial development began, coupled with a lack of knowledge of what research is required.

The chairman then asked "What are the expectations of research?".

It was felt that the major objective of research was the establishing of the feasibility of principles and theories, but there was a general belief that difficulties of knowing what research was required arose because of industries inability to outline what it needed, and often a lack of appreciation of the time scales involved in research initiatives. The academics present considered it important to emphasis that universities provided an equally important service for industry in the education of students.

It was pointed out that many managers believed their own engineers should be doing any research work which is required, and it was difficult to persuade these people who probably had no engineering background that research, development and maintenance required completely different skills.

Prof. D. Buchanan asked "Why, since industry had so many problems getting the research it required done, didn't it allow its own people to participate in the project?", this was echoed by many other academics. The industrialists replied that they were seldom invited and it seemed there was a problem over who should make the initial advance.

At this stage the chairman moved the discussion on a little asking, "What do we need and what is important to automate the textile industry?".

Several factors were seen as important; first an inspection system which does the work presently done by an operator. A general consensus from the academics said that the ability to achieve this was very dependent on what industry would invest in research and pay for the final product. A second factor was the development of new textile machinery, perhaps changing the design of some very common devices. This met with reserved approval from the industrialist who felt it was important to ensure that

NATO ASI Series, Vol. F 64
Sensory Robotics for the Handling of Limp Materials
Edited by P.M. Taylor

automation was gradual. Other important factors were sensors in a number of forms and better handling routines as a garment during production typically spends 80% of the time in various handling stages. It was then suggested that this list of inspection, machinery, sensors and handling was based on specific end-user needs and that a different list of priorities would be found among the textile machinery manufacturers and suppliers, where the major needs are, sensors, actuators, data processing and logistic process research.

At this point the chair asked "What would be on a research list of priorities?".

There was general agreement that industrial collaboration was vitally important with some industrialists of the opinion that no joint project should be undertaken unless the company involved was willing to invest. It was also felt that industry needed to be made more aware of the lead times involved in research and extrapolations should be made to allow for new technology which may be available in 5 or 10 years. The basic research needs were perceived to be in the areas of handling and inspection with AI being suggested as being useful. A major problem which was highlighted was the transfer of primary research from universities to industry. There was a feeling among the academics that the industrialists wanted a ready to use system, and at present the transfer was most easily achieved by researchers leaving universities to take a product to the market themselves. It was observed that, in line with the original specification that research was general and development specific, the list of requirements became more general in moving from the user, through the manufacturer to the university.

Finally the chairman asked, "Would this type of meeting have been held 10 or 15 years ago?".

The industrialists felt this would have been possible but would have involved individual companies with no 'enemy' present, while the university representatives expressed the opinion that 'mad' academics will always come together when a common area of work is found.

As a final comment Flavio asked if it would not have been an easier process to redesign clothing rather than trying to automate such a difficult product.

CHAPTER III

SENSING

2D VISION TECHNIQUES FOR THE HANDLING OF LIMP MATERIALS

Robert B. Kelley
Electrical, Computer, and Systems Engineering Department
Rensselaer Polytechnic Institute
Troy, NY 12180-3590, USA

Keywords: machine vision / robot vision / binary techniques / structured illumination / grey-level techniques / handling / in-process inspection.

Abstract — For automated robotic systems to handle limp materials, sensors are needed to determine the actual properties of the material in the supply, the material itself, and at the handling destination. This paper reviews 2D vision techniques with an aim toward satisfying the sensory goals associated with the handling and in-process inspection of limp materials. The 2D vision techniques are summarized in a discussion of their applicability in the context of typical questions asked of vision guided robotic systems.

I. INTRODUCTION

For most manufacturing processes, the automated handling of parts by a machine requires the parts to be properly positioned. In the case of handling limp materials, there is a significant challenge since such materials do not satisfy the usual assumption that the part is rigid. Of course, the rigid part assumption can be handled by stiffening the limp material or modeling the behavior of the material. For sensory robotic systems to handle limp materials, sensors must be used to determine the actual properties of the material supply, the material itself, and the handling destination. A machine vision system is a particularly suitable, general-purpose sensor. It is often the sensor of choice when there is a wide variety of parameters to be measured. Here, the focus is on the 2D vision techniques which are particularly appropriate to the sensory tasks that are associated with the robotic handling of limp materials.

To motivate the 2D vision techniques which are discussed, typical questions are listed in Table 1.1. These are posed to serve as an aid in identifying the various ways 2D vision might be used to supply the desired answers. The list of questions is aimed at accomplishing typical tasks, namely locating and acquiring a piece of limp material from a supply; determining the orientation, alignment and proper placement of a piece of material; and performing limited in-process inspection of limp materials which are handled by automatic systems.

NATO ASI Series, Vol. F 64
Sensory Robotics for the Handling of Limp Materials
Edited by P. M. Taylor

During the discussion which follows, it will be helpful to keep such questions in mind. First, some elementary 2D vision techniques are presented. These techniques are chosen with an aim toward satisfying the sensory goals which accomplish the typical tasks listed below. Then, the 2D vision techniques are summarized and discussed in the context of the typical questions posed for the handling of limp materials.

Table 1.1 Questions to motivate 2D vision technique selection

LOCATING AND ACQUIRING A PIECE OF LIMP MATERIAL FROM A SUPPLY.

0. *Where is the stack of parts?*

WHEN DEALING WITH A STACK OF PARTS:

1. *Where is the topmost part?*
2. *Where can the part be grasped?*

WHEN DEALING WITH ISOLATED PARTS:

3. *Is the part where it is expected to be?*
4. *How is the part resting?*

DETERMINING THE ORIENTATION, ALIGNMENT AND PROPER PLACEMENT OF A PIECE OF MATERIAL.

5. *What is the orientation of the part?*
6. *Is the part properly aligned?*
7. *How is the part resting?*

PERFORMING LIMITED IN-PROCESS INSPECTION OF LIMP MATERIALS.

8. *Is it the correct part?*
9. *Is it a good part?*

II. PREPROCESSING

This section begins the presentation of 2D vision techniques which can be applied to the handling of limp materials. No attempt is made to be exhaustive; rather, the purpose here is to direct attention to the variety of tools which are useful in this context.

Shading correction

Shading correction is used to transform a grey-level image into an image which approximates a uniformly illuminated one. For each pixel, the intensity measured by a camera is the product of the illumination incident on the reflecting surface, the surface reflectance, and the optical system of the camera. Further, the pixel value obtained depends on the transfer characteristic (intensity versus output) of the camera. In general, the transfer characteristic of the camera needs to be known to compensate for non-uniform illumination. Using a constant reference light source and varying the f-stops of the camera lens, the grey-level pixel value for a range of input intensities can be measured. This series of known aperture settings may be used to estimate the transfer characteristic.

Frequently, however, the transfer characteristic is modeled as a smooth exponential function of the form

$$g = g_o(I/I_o)^{\gamma}, \tag{2.1}$$

where g and g_o are observed and reference grey-levels, I and I_o are observed and reference intensities. (Several contemporary solid state cameras can be adjusted to have a γ parameter value of one.) For this transfer characteristic, shading correction can be obtained by scaling observed pixel values. The individual pixel correction scale factors are found as follows. The distribution of the non-uniform illumination of the scene is measured by viewing a calibration surface with a uniform, known reflectance. Usually the maximum over all the pixel values is taken as the reference grey-level g_o. With this information, observed grey-level values can be corrected to the values which uniform illumination would have given (within the quantization error). The correction factor is given by the ratio of the reference grey-level to the calibration grey-level for that pixel. That is,

$$k_{pixel} = g_{ref}/g_{pixel} \tag{2.2}$$

$$g_{corrected} = k_{pixel}\, g_{uncorrected}\,. \tag{2.3}$$

Automatic threshold selection

Binary images may be thought of as being a special subset of grey-level images. They can be obtained from grey-level images through the application of a two-valued transformation to each pixel. Usually a simple threshold is used for the entire image to segment it into object and background pixels. The pixels with grey-levels above the threshold are assigned one value and the rest the other value. If care is taken in obtaining approximately uniform illumination or if the contrast is strong enough (for example, when back lighting is used), then a single threshold value for the entire image may be satisfactory. Clearly, it is desirable to be able to use a single threshold value rather than to use extra processing effort to compensate for non-uniform illumination. The extra effort expended to obtain high quality images is rewarded by avoiding the repeated performance of extra preprocessing steps on every image. Corrections may be made to grey-level images for non-uniform illumination. Conventionally, the image is partitioned into non-overlapping regions. The threshold is adjusted in each region to approximate the result for a uniformly illuminated scene.

Automatic threshold selection requires that the image have good contrast and the regions be separable by means of a single threshold. Automatic selection of the threshold is based on the assumption that the histogram of each region generally has both object and background pixels. Various rules may be used to select the threshold for each region. If the histogram has two well-defined peaks with a flat valley, then some valley point is chosen as the threshold. Finding an extreme value of a criterion function is an objective way to select a *good* threshold which divides the histogram into two classes, object and background. An algorithm [Duda and Hart, 1] selects the threshold T which minimizes the criterion function $J_1(T)$ given by.

$$J_1(T) = \sum_{n=0}^{T} [h(n) - m_b(T)]^2 + \sum_{n=T+1}^{255} [h(n) - m_o(T)^2] , \qquad (2.4)$$

where T is the threshold, h(n) is the histogram function, and m_b and m_o are the mean values of the background and object pixels.

More recently, the following criterion function $J_2(T)$ was suggested for threshold selection [Otsu, 2].

$$J_2(T) = \frac{n_b(T)\, n_o(T)\, [m_b(T) - m_o(T)]^2}{[n_b(T) + n_o(T)]} , \qquad (2.5)$$

where T is the threshold, $n_b(T)$ and $n_o(T)$ are the number of background and object pixels, and m_b and m_o are the mean values of the background and object pixels. This is a smooth function and usually has a single maximum. When there is more than one maximum, then each maximum is examined to determine which maxima are associated with good thresholds. Two additional criteria, valley checks, are used [Kittler and Illingsworth, 3].

$$\sum_{j=-1}^{1} h(\tau + j) < \sum_{j=-1}^{1} h[m_b(\tau) + j] \qquad (2.6)$$

and,

$$\sum_{j=-1}^{1} h(\tau + j) < \sum_{j=-1}^{1} h[m_o(\tau) + j] , \qquad (2.7)$$

where $h(\cdot)$ is the histogram function, τ is the threshold corresponding to the local maximum, and m_b and m_o are the mean values of the background and object pixels for this threshold selection. These checks compare local averages of the number of pixel values about the chosen threshold and the class means, m_b and m_o. If there are two dominant grey-levels, both valley checks will be satisfied. For some images, only one of the check conditions may be satisfied; this is a signal that the underlying two dominant grey-level assumption may not be satisfied and the maximum found may not correspond to a good threshold.

Connectivity

When images are represented as a rectangular array of pixels, the determination of which regions are connected is complicated by the restriction that the sum of the connected neighbors for an object pixel and for a background pixel equals 12. A problem arises because it is natural to consider the 8 adjacent pixels of an object pixel as its connected neighbors. This then requires that background pixels be restricted to having just 4 adjacent pixels be designated as its connected neighbors. Clearly, the roles of the two kinds of pixels can be reversed. However, 8-connected objects have intuitive connection properties. (Some prefer to work with a symmetric definition where both object and background pixels have 6 neighbors [Agin, 4]; however, the selection of the 2 adjacent, non-connected pixels leads to other interpretation anomalies such as direction sensitive diagonal connections.) The different neighborhood definitions and a sample image array are shown in Figure 2.1.

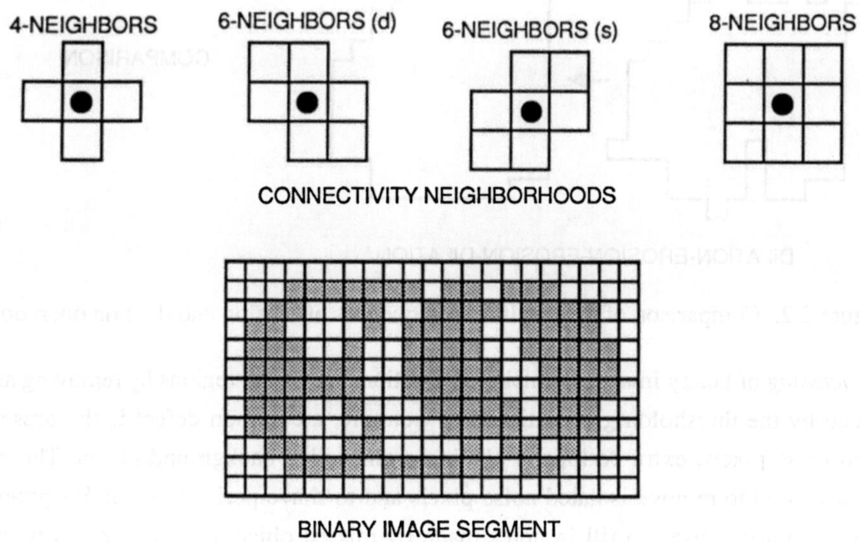

Figure 2.1. Connectivity neighborhoods and sample image array.

Erosion and dilation

Erosion and dilation are dual operations which are applied to regions in a binary image. The erosion operation removes pixels from an object region which are adjacent to a background pixel. That is, such object pixels are relabeled as background pixels. Similarly, the dilation operator relabels background pixels which are adjacent to object pixels. A comparison of erosion-dilation operations applied in both sequences is shown in Figure 2.2. The two operators are duals since eroding an object dilates the background and vice versa. These operators have many uses. They are used for postprocessing thresholded images, for generating templates, for emphasizing or isolating irregularities, etc.

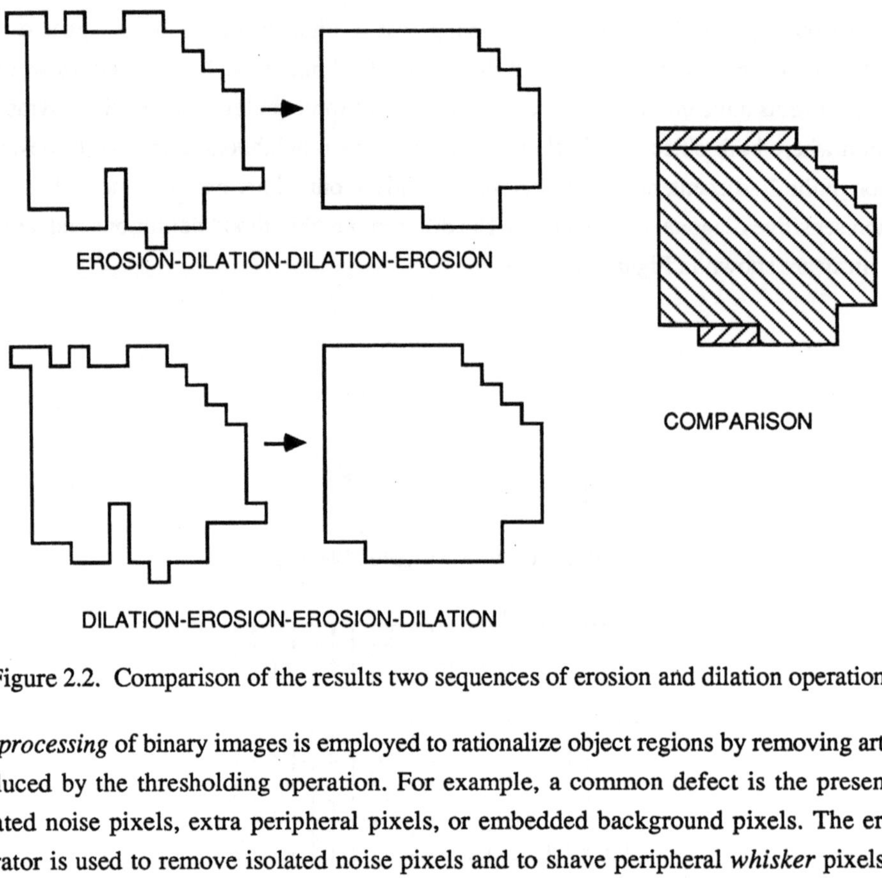

Figure 2.2. Comparison of the results two sequences of erosion and dilation operations.

Postprocessing of binary images is employed to rationalize object regions by removing artifacts produced by the thresholding operation. For example, a common defect is the presence of isolated noise pixels, extra peripheral pixels, or embedded background pixels. The erosion operator is used to remove isolated noise pixels and to shave peripheral *whisker* pixels. The dilation operator is used to fill in holes made by missed object pixels. For this purpose, a sequence of both operators is typically applied in performing the rationalization.

As defined above, clusters of noise pixels with a diameters up to 2 pixels and whiskers of 1 pixel length or 2 pixels width are removed in a single erosion operation; holes up to 2 pixels wide are filled in a single dilation operation. Thus the sequence of erosion, dilation, dilation, and erosion operations usually removes both kinds of defects and restores the object to its normal size. Alternatively, the sequence of dilation, erosion, erosion, and dilation operations can also be used. Differences between the actual performance of these two operator sequences arise when the initial operator changes the topology of the object region. Two object regions separated by no more than 2 pixels will become one region if the dilation operation is applied. Embedded holes within 2 pixels of the boundary of the object region will become joined to the background if the erosion operation is applied; if dilation is immediately applied, the gap may be closed again.

Masks or templates can be generated from the binary image of objects in a similar manner. In fact, the rationalized object region can serve as a mask to identify embedded holes and the like. To suppress spurious responses when using such a mask, extra dilation operations can be used to make the mask oversized. In this way small whiskers and small holes will be ignored. Difference images produced by these masks are useful in emphasizing irregularities and departures from the nominal object region topological characteristics.

III. GREY-LEVEL IMAGE SEGMENTATION

Grey-level images may be segmented by a variety of techniques. In this section, one geometry- and two intensity-based techniques are presented. In general, segmentation techniques have to be carefully selected to match the nature of the scene being segmented.

Structured illumination

Structured illumination is a technique which is sensitive to relative depth geometry. Thus, it can be used to segment a grey-level image by extracting silhouettes of objects as well as detecting surface irregularities on objects, such as wrinkles. The segmentation results in a binary image that is obtained by directly thresholding the grey-level image of the structured light pattern. The difference in grey-level images of the scene, with and without the structured light pattern, can be used to overcome interference from extraneous illumination. Segmentation is provided by noting the regions where there are deviations from the expected nominal structure of the projected pattern. The deviations are usually due to relative depth differences between regions of the scene. These deviations can be interpreted to obtain geometric information such as geometric edge location and orientation.

Consider a plane-of-light projector as shown in Figure 3.1. The nominal pattern for the intersection of the plane-of-light with a flat, reference surface is a straight line. Hence the nominal location of the line in the image corresponds to surfaces which are the same height as the reference. When the plane-of-light is inclined, relative surface height differences result in displacements of the line perpendicular to the direction of the nominal line pattern. The relative height difference is proportional to the displacement. (In fact, the displacement can be computed for each pixel of the line to provide a 3D slice through an object.)

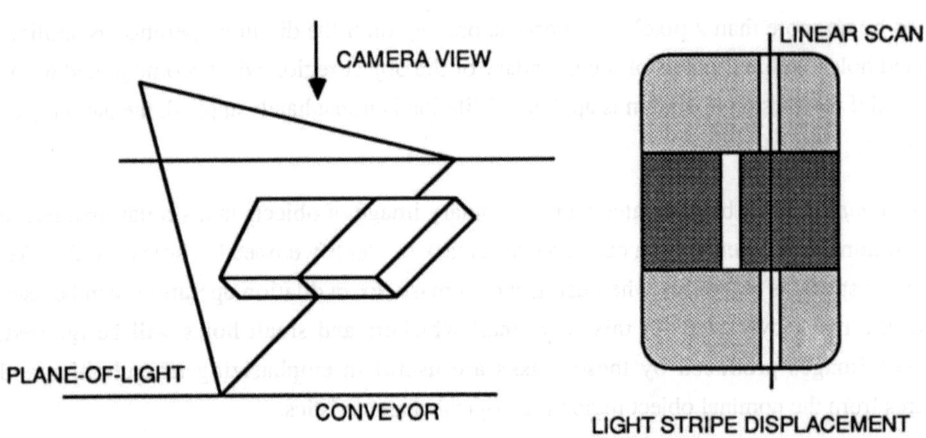

Figure 3.1. Plane-of-light silhouette extraction.

Silhouette extraction does not require any calculation. Rather the silhouette is obtained by simply noting where the nominal line pattern is broken. The number of projected plane-of-light measurements depends upon the application. A simple implementation which extracts the complete silhouette is obtained by moving the object relative to the line pattern. This can be realized by having the object on a conveyor, the projector on a track, or a robot arm carry all or part of the structured illumination and camera system. Multiple planes-of-light are a useful way to get more information from each image. A practical camera system for silhouette extraction would employ a linear array camera to provide faster scan times and simplify the silhouette extraction computation [Baird, 5].

Edge extraction

Edge extraction techniques associate large changes in grey-level values with geometric edges. Thus edge extraction is performed by means of edge detectors which are based on discrete approximations to directional derivative operators. Two orthogonal directional operators are used together as a gradient vector operator.

Popular edge detection operators include the 2x2 Roberts cross operator, and the 4-neighbor, 8-neighbor, and Sobel 3x3 operators. The Roberts cross operator effectively computes a diagonal difference which is centered at the 4-corner point, and is slightly more sensitive to higher spatial frequencies because of the size of the neighborhood [Roberts, 6]. The 4-neighbor operator computes pixel-centered orthogonal differences, has the least smoothing of the 3x3 operators. The 8-neighbor operator computes a weighted sum of orthogonal and both diagonal differences, has intermediate level of smoothing. The Sobel operator also computes a weighted sum of orthogonal and both diagonal differences with the orthogonal difference given a relative weight of 2, has the most smoothing and the smallest gradient direction error. The directional differences, gradient magnitude and direction expressions are listed in Table 3.1.

Table 3.1 Vector edge detection operators

Operator	Directional differences	Gradient magnitude	Gradient direction
Roberts cross	$d_1 = p_{i+1,j+1} - p_{ij}$ $d_2 = p_{i,j+1} - p_{i+1,j}$	$m_{ij} = \sqrt{d_1^2 + d_2^2}$	$\theta_{ij} = \arctan(d_2/d_1) - \pi/4$
4-neighbor	$d_i = p_{i+1,j} - p_{i-1,j}$ $d_j = p_{i,j+1} - p_{i,j-1}$	$m_{ij} = \sqrt{d_i^2 + d_j^2}$	$\theta_{ij} = \arctan(d_j/d_i)$
8-neighbor	$s_i = p_{i,j-1} + p_{i,j} + p_{i,j+1}$ $s_j = p_{i-1,j} + p_{i,j} + p_{i+1,j}$ $d_i = s_{i+1,j} - s_{i-1,j}$ $d_j = s_{i,j+1} - s_{i,j-1}$	$m_{ij} = \sqrt{d_i^2 + d_j^2}$	$\theta_{ij} = \arctan(d_j/d_i)$
Sobel	$s_i = p_{i,j-1} + 2p_{i,j} + p_{i,j+1}$ $s_j = p_{i-1,j} + 2p_{i,j} + p_{i+1,j}$ $d_i = s_{i+1,j} - s_{i-1,j}$ $d_j = s_{i,j+1} - s_{i,j-1}$	$m_{ij} = \sqrt{d_i^2 + d_j^2}$	$\theta_{ij} = \arctan(dj/di)$

Some prefer to use a scalar edge detection operator based on discrete approximations to the Laplacian as given in Table 3.2 or the Laplacian of a Gaussian smoothing filter [Horn, 7]. The argument is made that such operators improve noise suppression and edge localization properties. The tradeoff is between this performance improvement and the penalty represented by the loss of edge direction information. Commonly, an attempt is made to make up this loss by introducing external knowledge about the edge properties or some form of edge continuity assumption. In the extreme case, a scalar directional edge detector may be employed to obtain local edge direction information.

Table 3.2 Scalar edge detection operators: Laplacian approximation

Operator	Filter output
4-neighbor	$m_{ij} = 4p_{i,j} - (p_{i-1,j} + p_{i+1,j} + p_{i,j-1} + p_{i,j+1})$
8-neighbor	$m_{ij} = 8p_{i,j} - (p_{i-1,j-1} + p_{i-1,j} + p_{i-1,j+1} + p_{i,j-1} + p_{i,j+1} + p_{i+1,j-1} + p_{i+1,j}\ p_{i+1,j+1})$

Region growing

Region growing exploits small changes in grey-level values to aggregate adjacent pixels into common regions. Under uniform illumination, such regions would be associated with (almost) planar faces on the object. If only local differences are used, there can be a wide range of pixel values clustered together in the same region. On the other hand, if distant differences are used, the segmentation which is achieved often depends on the order in which the region growing was performed. Relaxation methods are employed to minimize this latter effect.

IV. BINARY IMAGE SEGMENTATION

Binary image segmentation requires the blobs in the scene to be identified or labeled. The highest level of description of the segmentation is given by the connected components tree. Isolated blobs can then be described in terms of their contours or by their moments (Section V).

Connected components

This analysis produces a connection tree which identifies the objects in the image, the holes entirely enclosed by each object, the objects found inside each hole, and so forth. As a byproduct, connected component analysis usually produces parameters which are specific to each element in the connectivity tree: such as its area, centroid, moments and axes of inertia, etc. These parameters are treated in Section V. The position and orientation of any blob is easily computed from such parameters.

Contour extraction

A search technique such as contour extraction may be used to segment a binary image into distinct blobs. Contour extraction requires a starting point on the edge of the blob. For sequential computers, an easy way to locate a starting point on the blob edge is to scan along a row or column until the edge of a blob is crossed. If a reference point is known to be interior to the blob, then starting the scan for the edge from the reference point is efficient. A standard *edge following bug* is used to extract a contour description of the blob. A simple edge following bug traces the contour in a fixed direction [Ballard and Brown, 8]. At each step, it examines its neighbors in order while keeping the interior on the same side as it traces the contour. If is outside the blob, it searches for the first neighbor in the forward direction which is in the blob and conversely. This bug will fail on thin, hair-like blobs; using two bugs which trace the contour from the left side and from the right side solves this problem [Dessimoz,9]. The extracted contour description allows the position and orientation of the blob to be determined by techniques treated in the next section.

V. Implicit Shape Features

These features are obtained by using information which is not constrained to be local to the immediate neighbor of any particular pixel. Such features include moments, transforms, topological properties, and interfeature relationships. These features may be used to describe the shape, position and orientation, or other patterns for purposes of location or recognition.

Shape description

To describe the shape of an object, the object must be distinct from the other objects in the image; that is, the image must be segmented and the object isolated from the rest of the scene. Hence, for this discussion, it is assumed that there is only one object present. Shape may be described in terms of implicit parameters or explicit geometry-based parameters such as boundary curvature discontinuities (see Section VI).

Implicit parameter descriptions are those as given by the 2D mechanical (or statistical) moments, or by Fourier transform coefficients for the uniformly weighted pixels, for example. For single valued objects as obtained from a binary image, all the shape information is contained in its boundary relative to the background (and embedded holes). Because the interior pixels of the object dominate these parameters, some researchers treat the interior pixels as background pixels and give them zero weight. This increases the sensitivity of such parameter calculations to perturbations in the number and location of the peripheral pixels.

The lower order moments are also generally used in the preparatory steps of other computations. The zero-order moment gives the area of the object. The first-order moments give the coordinates of the center of mass, the centroid – an object reference position point. The second-order central moments are computed with respect to the centroid as the coordinate origin and provide a reference orientation direction (with an ambiguity of π).

The mechanical/statistical moments, up to the second-order, are computed as follows. Let p_{ij} be the pixels with non-zero weights. Then, the area is given by

$$m_{00} = \Sigma_i \Sigma_j p_{ij} , \tag{5.1}$$

the centroid (m_{10}, m_{01}) by

$$\begin{aligned} m_{10} &= (1/m_{00}) \Sigma_i \Sigma_j \, i \, p_{ij} , \\ m_{01} &= (1/m_{00}) \Sigma_i \Sigma_j \, j \, p_{ij} , \end{aligned} \tag{5.2}$$

and the central moments of inertia by

$$m'_{20} = (1/m_{00})\, \Sigma_i \Sigma_j (i - m_{10})^2 p_{ij} ,$$
$$m'_{11} = (1/m_{00})\, \Sigma_i \Sigma_j (i - m_{10})(j - m_{01}) p_{ij} ,$$
$$m'_{02} = (1/m_{00})\, \Sigma_i \Sigma_j (j - m_{01})^2 p_{ij} . \quad (5.3)$$

The direction of the axis of the least moment of inertia is often used as a modulo-π orientation direction reference. Higher-order moments can be computed in a similar fashion. Since the third-order moments are modulo-$2\pi/3$, they can be used to disambiguate the polarity assignment for the direction of the axis of least moment of inertia.

Polar Moments

Some rotational invariant moment measures (polar moments) can be computed directly from the mechanical moments given above. Define the polar moment magnitude as

$$M_{ij} = \sqrt{C_{ij}^2 + S_{ij}^2} \quad (5.4)$$

where the C_{ij} and S_{ij} are related to the m_{ij} by the following relationships:

$$
\begin{array}{llll}
C_{00} = m_{00} & S_{00} = 0 & C'_{31} = m'_{30} + m'_{12} & S'_{31} = m'_{03} + m'_{21} \\
C_{11} = m_{10} & S_{11} = m_{01} & C'_{22} = m'_{20} - m'_{02} & S'_{22} = 2\, m'_{11} \\
C'_{20} = m'_{20} + m'_{20} & S'_{20} = 0 & C'_{33} = m'_{30} - 3\, m'_{13} & S'_{00} = 3\, m'_{00} - m'_{00}
\end{array} \quad (5.5)
$$

Contrast invariance is provided by dividing the higher-order moments by the area, m_{00}. *Image size invariance* is provided dividing the polar moment M_{ij} by R^{2i}, where R is the radius of gyration computed as

$$R = \sqrt{(m'_{20} - m'_{02})/m_{00}} . \quad (5.6)$$

SRI-type geometry-based parameters

Originally developed at SRI in the 1970s, such features (frequently referred to as SRI features) [4] have been used to describe not only the object's shape but also its topology. Efficient algorithms were developed in the early 1970s, principally at SRI International. These algorithms produce explicit geometry-based descriptive parameters which are primarily used for feature recognition. SRI used special purpose vision capture hardware which directly output a run length representation of the binary image of the scene; this was further processed by a minicomputer. Similar vision systems were marketed in the US and Europe in the late 1970s. A *partial* list of SRI-type features is given in Table 5.1.

Table 5.1 Partial list of SRI-type features

The surface area of the object;
The length of the enclosing perimeter;
The total area of all holes;
The maximum distance measured from the centroid to perimeter pixels;
The minimum distance measured from the centroid to perimeter pixels;
The root-mean-square distance measured from the centroid to perimeter pixels;
The angle(s) between the rays from the centroid to the maximum distance perimeter pixel(s);
The angle(s) between the rays from the centroid to the minimum distance perimeter pixel(s)
The angle to the centroid of embedded holes, corner pixels, and other features, measured relative to the direction of the ray from the centroid through the maximum perimeter pixel; and
The compactness of the blob, defined as the ratio of the square of the perimeter length to the area with a minimum value of 4π for a circle.

VI. Explicit Shape Features

The shape of the perimeter of an object can be represented by a sequence of the directions of steps taken as the entire perimeter is traversed. The direction sequence is cyclic. If the starting pixel is chosen as the maximum distance pixel, for example, the sequence of relative direction changes is (ideally) unique and rotationally invariant. Because of the spatial quantization effect of array cameras, the rotational invariance is only approximate. However, if the step distances are taken into account and the Fourier Series coefficients are computed, then removing the higher-order Fourier coefficients of the relative direction sequence (that is, the sequence is smoothed by a non-causal low-pass filter) gives a slightly rounded but more rotationally insensitive representation [9].

Chain code

The chain code is a representation of the perimeter of an object in terms of the step direction as the perimeter is traced one pixel at a time. Traditionally, the directions are taken to be either the 4-neighbor or 8-neighbor directions. If the step neighborhood is extended beyond the 3x3 neighborhood, somewhat better direction angular resolution can be obtained. Alternatively, the sequence of step directions can be smoothed over a window covering several steps. In all cases, it is important to take the pixel-to-pixel distances into account or suffer severe distortions when the perimeter is rotated on the rectangular image lattice. The chain code also exhibits a discontinuity if there are 2π direction changes along the perimeter. Modulo-2π reduction must be used to handle rotations. Differential direction chain codes are well suited for a rotation invariant shape representation.

Curve fitting

In addition, a plot of the differential directions versus the step distance around the perimeter provides an easy way to extract a constant curvature, or *concurve*, representation of the perimeter [Perkins, 10]. Thus the concurves represent the perimeter as connected segments of straight lines and circular arcs. In the differential direction vs. step distance plane, concurves form horizontal lines. Hence, the best horizontal line fit automatically finds the break points for the concurve segments.

Line fitting

Fitting a line to a set of scalar point locations is usually done according to a best-fit technique. The eigenvector with the larger eigenvalue which passes through the centroid of the cluster of points is frequently computed and gives the least square error line fit. If the points carry direction information, then the direction histogram can provide not only the direction of the best fitting line, but also identify those points which are farthest from the line. Elimination of such outliers usually improves the quality of the line fit. Hough transform techniques which are also used in scalar point straight line fitting are made greatly more efficient by the inclusion of line direction information to limit the dispersion of the Hough accumulator entries [8].

VII. Summary Discussion and Conclusion

There are a large number of techniques which might be used applied to the handling of limp materials. A brief review of some promising techniques has been given. The short list of application related questions from Section I (repeated here) gives context to the issues discussed.

0. Where is the stack of parts? This question has a trivial answer if the parts are where they are expected to be. If not, prior knowledge about the supply stack should provide a guide to the selection of appropriate 2D vision techniques. Structured light appears to be the most universal, reliable technique to answer such questions.

1. Where is the topmost part on a stack of parts? Structured light again provides the most reliable way to locate the topmost layer of fabric or leather in a stack. Otherwise, grey-level edges can be used to extract an estimate of the geometric edge locations. Binary blob extraction might also suffice.

2. *Where can the topmost part be grasped?* The answer to the question depends strongly on the gripper selected; the part itself can constrain the available holdsites. Surface attachment grippers, such as needle point devices, require a site which is virtually flat and of sufficient size. Structured light and binary blob extraction are the most promising approaches. Clamping grippers need to locate at least one edge of the topmost part. Structured light is indicated again, with grey-level edge detection a reasonable alternative.

3. *Is the isolated part where is is expected to be?* This question requires the part position to be be checked. A binary image blob centroid location is sufficient. Similar computations can be performed on contour or edge data.

4. *How is the isolated part resting?* This requires the part orientation to be determined. Again blob moments, especially polar moments, are usually sufficient. Explicit geometric features such as corner and hole locations can provide robust orientation estimates.

Position and orientation corrections. The displacement between the actual and nominal feature centers or centroids can be used, together with robot eye-to-hand calibration, to update the nominal location commands in the robot control program. Any rotational correction which may be needed is obtained by applying the camera horizontal and vertical scale factors to the components of the actual and nominal reference direction vectors. Rotational corrections can also be computed by projecting the image coordinates onto a robot coordinate system calibration plane. (Note that angular measurements are sensitive to image coordinate scale factors, but are independent of the origin of the coordinate system employed.)

5. *What is the orientation of the part?* Often it is prudent to verify the orientation of a part while it is in the grasp of the gripper. Parts sometimes suffer from minor slippage when initially grasped. Question 4 techniques can be applied to this case. The presence of the gripper in the image needed to be accommodated.

6. *Is the part properly aligned?* This is also a verification step to check the part in the gripper and is approached in a manner similar to that in question 3.

7. *How is the part resting?* This is a verification of the quality of the placement of the part at its target location. It includes position, orientation and alignment with other parts. If questions 5 and 6 have been answered, then only the relative alignment needs to be verified. Structured light is the indicated approach. Grey-level edge detection is a reasonable alternative to detect misaligned edges. Prior knowledge needs to be exploited to simplify the interpretation of the possibly cluttered scene image.

8. *Is it the correct part?* There is always a need to verify the correctness of the part being handled. This is simplest when the part is isolated or being held in the gripper. If the question is asked in response to an alignment failure, then the task is more difficult. However, the extraction of the most salient features can simplify getting the answer.

9. *Is it a good part?* This is a variation of question 8 which focuses on the quality of the correct part or on its placement. For example, surface irregularities such as wrinkles can be detected by using structured light patterns with an area array camera. Departures from the nominal 2D pattern provide the means to locate surface smoothness irregularities. By isolating the departures from the nominal pattern, the direction of the wrinkle and, hence, the direction to apply the smoothing action can be determined. Small wrinkles could be ignored, if appropriate, by noting the magnitude of the pattern deviation.

Table 7.1. Suggested techniques versus typical question

	Salient features	Blob/polar moments	Contour/edge data	Binary blob data	Grey-level edge data	Structured lighting
0. Where is the stack of parts?	.	.	.	.	.	X
1. Where is the topmost part on the stack?	.	.	.	X	X	X
2. Where can the topmost part be grasped?						
surface gripper	.	.	.	X	.	X
clamping gripper	.	.	.	.	X	X
3. Is the isolated part where it is expected to be?	.	.	.	X	X	X
4. How is the isolated part resting?	X	X	.	.	.	X
5. What is the orientation of the held part	X*	X*	.	.	.	.
6. Is the held part properly alignmented?	.	.	X*	X*	.	.
7. How is the part resting?	.	.	.	.	X	X
8. Is it the correct part?	X	X	.	.	.	.
9. Is the part good?	X	X	.	X	X	X

* The presence of the gripper in the scene must be accommodated.

Conclusion

Of the many 2D vision techniques which are applicable to the vision guided handling of limp materials, the use of structured light patterns, binary blob analysis, and simple edge extraction techniques seem to be able to provide answers to the typical question encountered in applications. The suggested techniques are listed for each question in Table 7.1. Other, more sophisticated techniques may be required to handle special cases which fall outside the simplified scenario treated in this paper.

REFERENCES

1. Duda, R. and P. Hart: Pattern classification and scene analysis: New York: Wiley-Interscience 1973

2. Otsu, N.: A threshold selection method from gray-level histograms: IEEE Trans. Systems, Man, Cybernetics, **SMC-9**, 62-65 (1979)

3. Kittler, J. and J. Illingsworth: On the threshold selection using clustering criteria: IEEE Trans. Systems, Man, Cybernetics, **SMC-15**, 652-655 (1985)

4. Agin, G.: Vision systems. In: Handbook of industrial robotics, S. Nof (ed.), pp. 231-261: New York: Wiley 1986

5. Baird, M.: Image segmentation techniques for locating automotive parts on a conveyor belt: 5th Intl. Joint Conf. on Artificial Intelligence, Cambridge, MA: 694-695 (1977)

6. Roberts, L.: Machine perception of three-dimensional solids. In: Optical and electro-optical information processing, J. Tippett et al. (eds.): Cambridge, MA: MIT Press 1965: 159-197

7. Horn, B.: Robot Vision: New York: Wiley 1986

8. Ballard, D. and C. Brown: Computer vision: New York: Prentice-Hall 1982

9. Dessimoz, J-D.: Traitment des contours en reconnaissance de forms visuelles (in French): Switzerland: Ecole Polytechnique Federale de Lausanne, Ph.D thesis 1980

10. Perkins, W.: A model-based vision system for industrial parts: IEEE Trans. Computers, **C-27**, 126-143 (1978)

3-D SENSING FOR ROBOT VISION

J. Domey, M. Rioux and F. Blais
Photonics and Sensors Section
Division of Electrical Engineering
National Research Council Canada
Ottawa, Canada
K1A 0R8

ABSTRACT

Machine vision is becoming an integral part of advanced manufacturing systems. Most of the systems presently in use are 2-D. Future vision systems will involve a more direct measure of 3-D information. This paper reviews some popular techniques used in 3-D sensing for robot vision. Basic theory of triangulation and laser beam propagation properties are presented along with typical industrial applications.

1. INTRODUCTION

Why 3-D sensing? Mainly because 2-D pictures are projections of 3-D scenes in a 2-D space. For very constrained situations where most objects of interest are flat and have few stable positions, 2-D imaging can be tuned to extract geometrical features with enough accuracy for manipulation and inspection. Otherwise, the 2-D projections of the 3-D objects are ambiguous. They are subject to modifications with any change of the illumination source, its spectral characteristics, or its orientation relative to the scene and/or the camera. The projected contours in 2-D imaging also change with a change in the oriention of the object, a translation in the field of view, and with position of nearby objects which scatter incident light on the object of interest.

*NRC No. 29409

NATO ASI Series, Vol. F 64
Sensory Robotics for the Handling of Limp Materials
Edited by P. M. Taylor

A change in distance from the camera produces a scaling effect that must be taken into account at the image processing stage. As the processing is based on intensity data, in addition to the geometrical effects, the surface properties such as color and texture will modify the value of each pixel and their relationship to each other.

What motivates research in this direction is that the eye-brain system in humans has resolved the difficulty of 3-D reconstruction and as a consequence researchers know that it is possible to make sense from 2-D imagery. How is it done? Answering this question will require many more years of research. But it must be remembered that the task is complex.

In contrast, 3-D imagery keeps the invariance properties of 3-D objects. The relationship between pixels is invariant with rotation and translation. With structured light approaches, for example, the 3-D data are insensitive to background illumination and to surface color and texture of the object. These invariant properties, combined with the absence of scaling effects, make the image processing tasks easier. Although it seems more difficult, computer processing of 3-D data is greatly simplified.

As an illustration, Figure 1 shows 3-D and 2-D pictures of human faces. The illustrations in (b) are the 3-D data where the range (or elevation) is coded in grey level for display. These pictures represent the x, y, z coordinates of the surface topography of the faces. Figure 1(a) is an intensity picture (essentially a normal photograph) of the same subjects (x, y, I). The eye-brain system has learned to make 3-D reconstruction from this type of representation, but, if the mode of representation is changed (as illustrated in (c) and (d)), it is interesting to note that now, in the case of an isometric representation, it is the x, y, z display that makes sense to the human observer. The isometric view of the x, y, I shows the complexity of the processing required to recover the geometry.

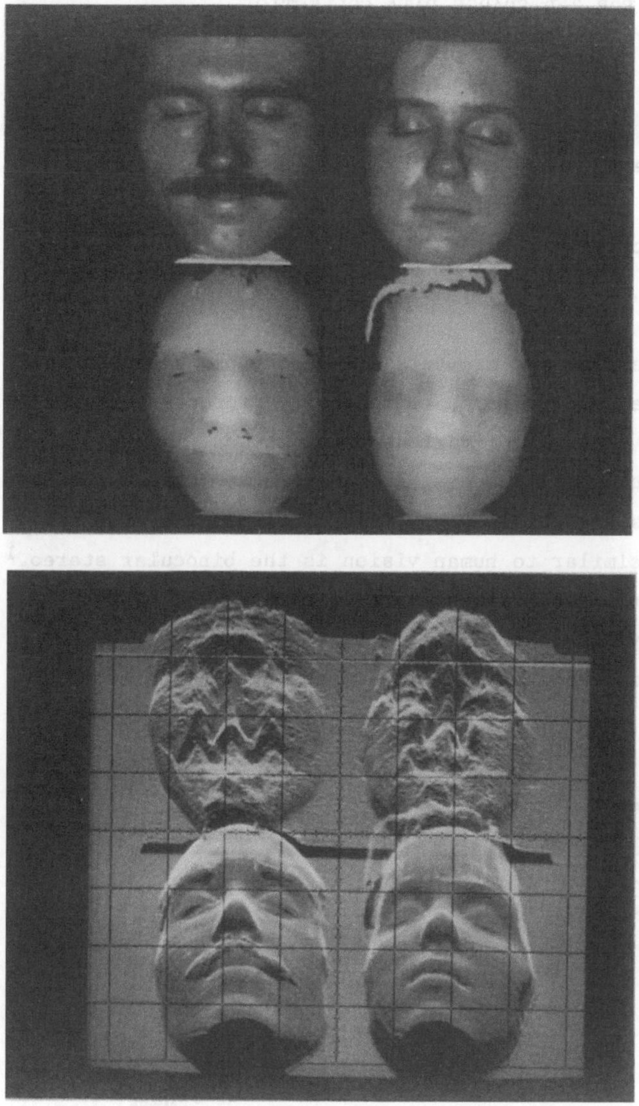

Figure 1. Human faces recording: (a) x, y, I or the equivalent of a photograph, (b) x, y, z which is a 3-D measurement of the subject's face, (c) an isometric view of the data displayed in (a), (d) the same for the 3-D recording.

2. BASIC TECHNIQUES FOR 3-D SPARSE DATA COLLECTION

A large number of configurations are used to extract 3-D coordinates from a scene. They can be grouped in two categories:

(a) sparse data collection and
(b) imaging devices.

Most sparse data collection arrangements are variations based on human binocular vision. They use two or more intensity pictures (usually in ambient light) of the scene along various orientations. Large variations of intensities are detected and correlated between the views; 3-D coordinates are obtained by triangulation. The simplest arrangement which is similar to human vision is the binocular stereo.[1] Figure 2 shows the geometrical arrangement which uses one or two cameras. The objective is to take two pictures of a scene from two different views. This can be accomplished using one moving camera

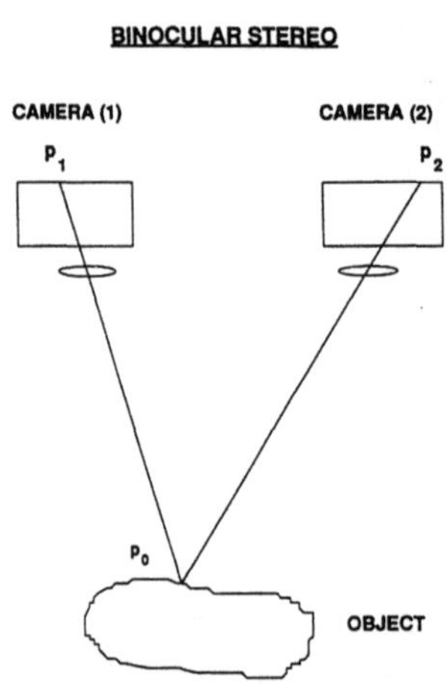

Figure 2. A geometry inspired by human vision.

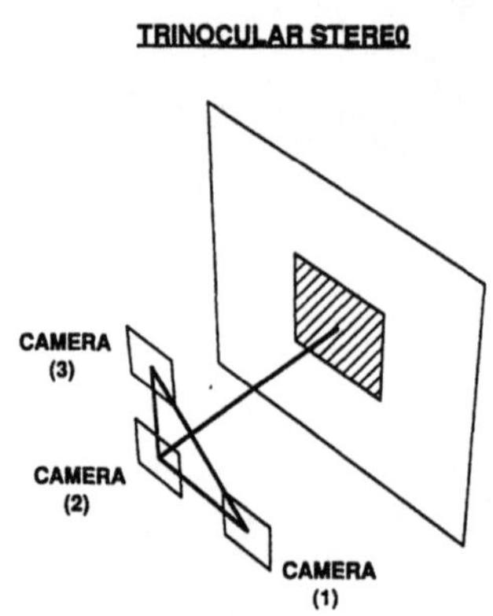

Figure 3. Using a third eye to improve matching.

(when the scene is stationary) or two fixed cameras which take the two views simultaneously. Figure 2 shows how a particular point in the scene is imaged within the cameras. P_0 is imaged in P_1 for camera 1 and is imaged in P_2 for camera 2. The whole exercise is to find, without ambiguities, the pair of points originating from a single point in the scene. The searching strategy is based on epipolar lines on which correspondence occurs. When correspondence has been established, then the coordinates are obtained by triangulation (a detailed computation is given in Section 4.)

In the trinocular arrangement[2] (Figure 3) a third view is taken at an orthogonal direction in order to increase the amount of correspondence. Indeed, horizontal edges cannot be measured using only cameras 1 and 2. Similarly, vertical edges cannot be measured using only 2 and 3. By making correspondence using the three pairs, a better matching is obtained and the redundancy (of the geometry) allows fewer errors to filtrate.

Another stereo variation[3] is shown in Figure 4. In this case the two or more views are taken along the optical axis of the camera. The correspondence occurs on radial lines. This approach has obvious advantages for vehicle guidance.

Generally speaking the stereo approaches which give sparse 3-D data require large amounts of computer time, but the data have relevant geometrical features for image interpretation. Indeed, most of the large variations in intensity come from edges that can be used to reconstruct or recognize an object.

When surface data are needed then 3-D imaging techniques are more appropriate.

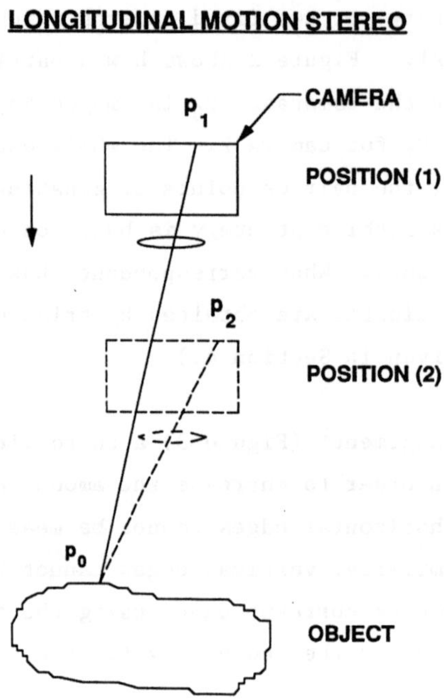

Figure 4. Illustration of a motion stereo approach.

3. BASIC TECHNIQUES FOR 3-D IMAGING

To obtain range information on areas of a scene which have uniform reflectivity, a number of active techniques have been developed. Basically a known pattern of light is projected onto the scene and an appropriate sensor used for its detection. The pattern can be spatial or temporal. These are often called structured light approaches.

Figure 5 illustrates a radar-type arrangement[4] where a modulated laser beam is sent to a scene to measure the range between the source and the surface element scattering back the incident beam. In this configuration the laser beam is well collimated and a scanning device covers the desired field of view. For each pixel a time interval is measured which corresponds to the range z using the following expression;

$$z = ct/2 \qquad (1)$$

where c is the speed of light and t the transit time.

A number of modulation techniques have been used. The simplest is to send a short impulse of light (Figure 5a). Amplitude and frequency modulation are also used[5] in order to improve accuracy by signal averaging (Figure 5b). In this case a phase shift is measured and multiple frequency modulation removes the 2π ambiguity.

One of the advantages of active techniques is the high density of pixel measurements which allows surface curvature measurements and segmentation.

Figure 6 illustrates the basic components used in interferometry type arrangements. A spatial modulation is imposed on the source and projected onto the scene. A camera which contains a reference pattern similar to the projected one takes a picture of the scene. The resulting image has a Moiré pattern from which the surface profiles are extracted. Again, a very large number of variations have been developed. An interesting arrangement[6] uses multiple-frame phase-shifted Moiré. A sequence of images is taken with the grid located at different positions. As an example, a first image is taken

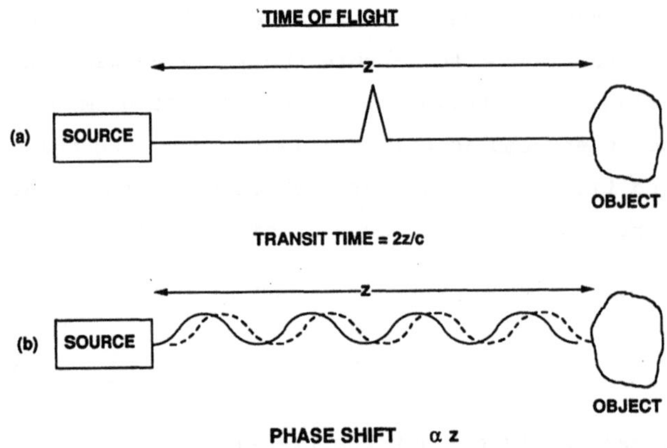

Figure 5. Radar type imaging device. In (a) a pulse is sent to the scene; in (b) a modulation is used.

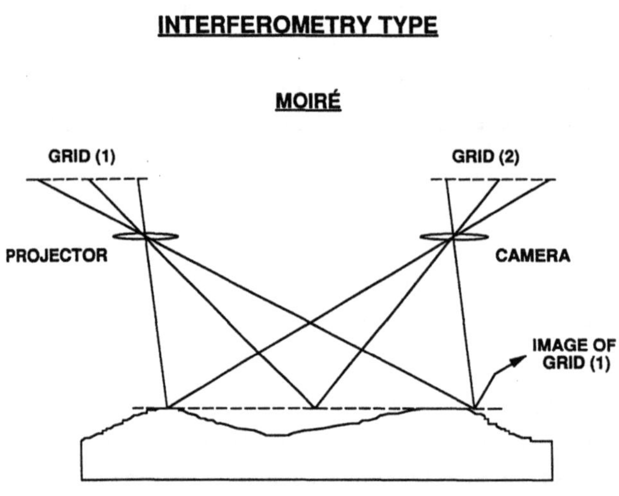

Figure 6. In a Moiré arrangement, a spatial modulation records surface height variations.

with an arrangement as shown in Figure 6, then a second image is taken after precisely moving the grid 1 by an amount which corresponds to a fraction of the grating period. N frames are taken similarly. Phase measurements for each and every pixel in the field of view are made. As in radar type devices the phase measurements are converted to z coordinates using a simple relationship.

Another popular method for obtaining dense 3-D data measurements is by triangulation using laser beam projection. The following is a detailed analysis of this method that can also be applied to all off-axis methods, including passive stereo arrangement.

4. THEORY

Within certain limits, all types of triangulation-based 3-D range finders can be analyzed using standard goemetrical optics. While not complete, this analysis provides a good background which permits the evaluation of the possible or expected performances of a range sensor. From these basic considerations, other aspects of the real application such as the practical implementation of the specific vision system, its acquisition speed, sensor size, stability, complexity, maintenance, and cost can be considered.

4.1 SINGLE POINT OPTICAL TRIANGULATION

The basic geometrical principle of optical triangulation is represented in Figure 7. The light beam generated by the laser is deflected by a mirror and scanned on the object. A camera, consisting of a lens and a position-sensitive photodetector, measures the location of the image of the illuminated point on the object. By simple trigonometry the three-dimensional coordinates of that point are calculated.

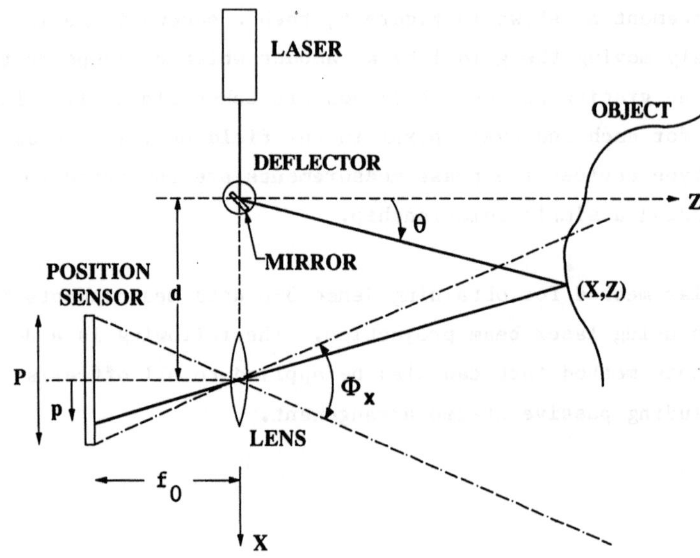

Figure 7. Basic geometry for optical triangulation.

From Figure 7 we have

$$z = df_0/(p + f_0 \tan \theta) \tag{2}$$

and

$$x = z \tan \theta \tag{3}$$

where p is the position of the imaged spot on the photodetector, x and z are the 3-D locations of the illuminated point on the object, θ is the deflection angle of the laser beam, d is the separation between the lens and the laser source, and

$$f_0 = \ell f/(\ell - f) \tag{4}$$

where f is the focal length of the lens and ℓ the distance to the object plane. Assuming that $\ell \approx \infty$ then $f_0 \simeq f$ and equation (2) becomes

$$z = df/(p + f \tan \theta) \tag{5}$$

Using this geometry, the absolute error on the distance Δz is

$$\Delta z = \left| \frac{z^2}{fd} \right| \Delta p + \left| \frac{z^2}{d \cos^2 \theta} \right| \Delta\theta \tag{6}$$

where Δp and $\Delta\theta$ are the measurement errors on p and θ, respectively. The accuracy $(1/\Delta z)$ of the range sensor is therefore directly proportional to the separation between the laser and the position sensor, and the focal length of the lens, but inversely proportional to the square of the distance. The accuracy is also closely related to the measurement angle θ, varying as the square of the distance.

With z, Δx can be approximated independently of the triangulation method

$$\Delta x = | \tan \theta | \Delta z + \left| \frac{z}{\cos^2 \theta} \right| \Delta\theta \tag{7}$$

The gain of a triangulation system is a useful parameter defined as

$$M_{3D} = \frac{\Delta p}{\Delta z} = \frac{fd}{z^2} \tag{8}$$

which indicates that for a given distance z and position sensor resolution Δp, the acccuracy of the sensor is directly related to the focal length of the lens of the position sensor f and the separation between the laser source and the position sensor d. Unfortunately, f and d cannot be made as large as desired: d is limited by the mechanical structure of the optical set-up and by the shadow effects. As can be seen in Figure 8, d must be kept small to minimize these effects.

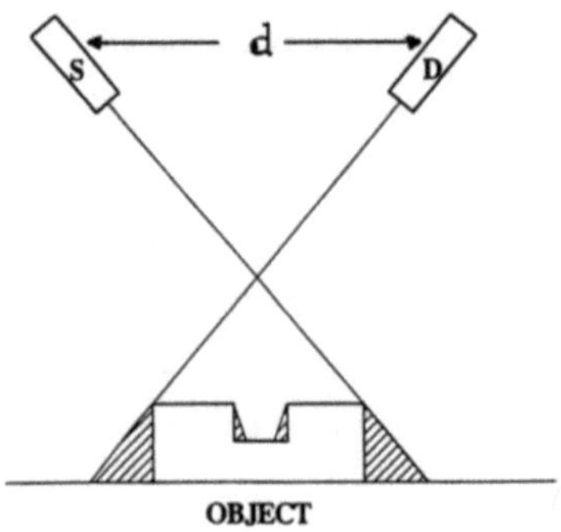

Figure 8. Shadow effects.

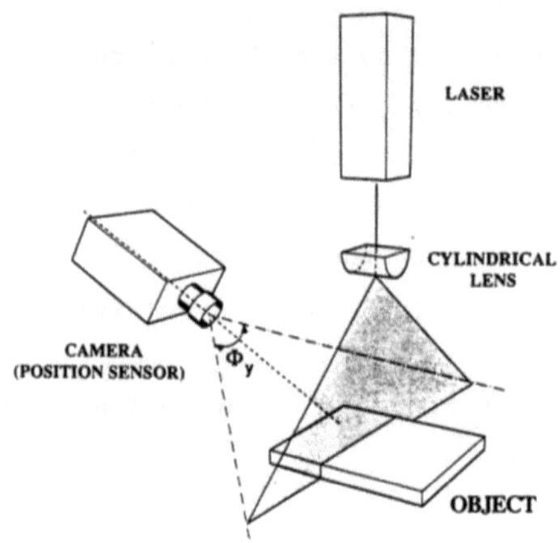

Figure 9. Plane of light projection.

In conventional triangulation geometry the field of view of the sensor, when the light beam can be scanned over the whole field, is approximated by

$$\phi_x = 2 \tan^{-1} (P/2f_0) \simeq 2 \tan^{-1} (P/2f) \qquad (9)$$

where P is the linear dimension of the position detector. Therefore, in the conventional setup, a compromise between the field of view and the accuracy of the 3-D measurement must always be considered. From equations (8) and (9)

$$\Delta z = (2z^2/d)\left(\tan(\phi_x/2)\right)(\Delta p/P) \qquad (10)$$

$$\Delta z = (zX/d)(\Delta p/P) \qquad (11)$$

where X is the desired measurement length at distance z, and the ratio $\Delta p/P$ is the resolution of the position sensor, normally a fixed quantity (e.g., the number of resolvable elements on a CDD) which confirms previous discussion of compromise between the field of view, accuracy of the 3-D measurements, and this shadow effect.

4.2 SINGLE AND MULTIPLE LINES

The second classical approach consists of projecting a laser line onto the object and measuring the 3-D coordinates of the points on the projected line using a bidimensional position sensor (normally a CCD camera) as shown in Figure 9. The line is generated by spreading the laser beam with a cylindrical lens along the y axis of Figure 7 (perpendicular to the xz plane)

There are certain advantages to using this approach. Because no moving parts are involved in the triangulation process, the method is simple, inexpensive, and robust. The main disadvantage, however, is due to the compromise between field of view and accuracy on the z coordinate as discussed previously but here given by

$$\Delta z = (2z^2/d)\left(\tan(\phi y/2)\right)(\Delta p/P') \qquad (12)$$
$$= (zY/d)(\Delta p/P')$$

where Y is the desired length of the field of view and P' the overall dimension of the position sensor parallel to the projected line. As in the classical triangulation set-up, good accuracy in the distance measurement will introduce a poor field of view, and vice versa.

The ambient light immunity is also reduced because of the dispersion of the laser light on the line. Longer integration time of the light on the position sensor will be required to obtain the same video signal and consequently the ambient light (or noise) signal will be seriously increased. The immunity is also reduced because of the large field of view required to cover the whole scene.

When high immunity to ambient lighting is not required, this type of approach has interesting properties for many robotic applications where high precision in the distance measurement is not required as is the case for the control of manipulators. Such a 3-D sensor can be easily mounted on a robot arm. Added to the low cost, simplicity, and compactness of this type of 3-D sensor, the simultaneous acquisition of a complete profile simplifies the subsequent processing of the data when the object is moving rapidly and nonuniformly, as may be the case when dynamic control of robots is required (e.g., grasping of objects). Contrary to the flying spot 3-D technique, there is no need to compensate for the relative lateral or transversal displacement of the object with respect to the sensor.

Moving the laser line, moving the camera head, or projecting multiple stripes of lines are the usual ways of increasing the density of measurements. Projection of multiple lines has the advantage of not introducing any mechanical moving parts or displacement. The method also has the advantage of increasing the relative speed of simultaneous measurement and automatically correlating all 3-D displacements of the object (a single line correlates only depth and lateral movements).

4.3 ARRAYS OF DOTS AND ENCODED PATTERNS

When many lines are used to illuminate the scene and only one image is used, ambiguities can arise whenever depth discontinuities are present. When the density of projected lines is high it is not clear which detected line corresponds to which projected line. Instead of using N images of M projected lines to obtain MN lines, several methods have been described using spacially encoded projected patterns to reduce the number of images to be processed to $\log_2 N$. The idea consists mainly of encoding the projected image following a predefined code, generally a binary or Gray code. Some systems have been demonstrated using arrays of dots, encoded lines, and encoded patterns.

Vuylsteke and Oosterlinck[7] have developed an interesting binary coded scheme that requires only one projected binary pattern, therefore reducing the complexity of the projector and the disadvantage of multiple images. The basic idea is to use a single binary coded projection mask in which each local neighborhood of indicated points reflects a specific binary signature. For low profile fast moving objects, this approach has many interesting advantages.

4.4 TRIANGULATION THROUGH THE LENS

Instead of using the position of the projector to compute the 3-D coordinates of the point on the object, two or more views can be utilized. One possibilility is to involve two or more cameras during the triangulation as discussed previously. Another approach consists of using the focusing properties of a standard camera lens. Using a double aperture mask in front of a single CCD camera provides a very

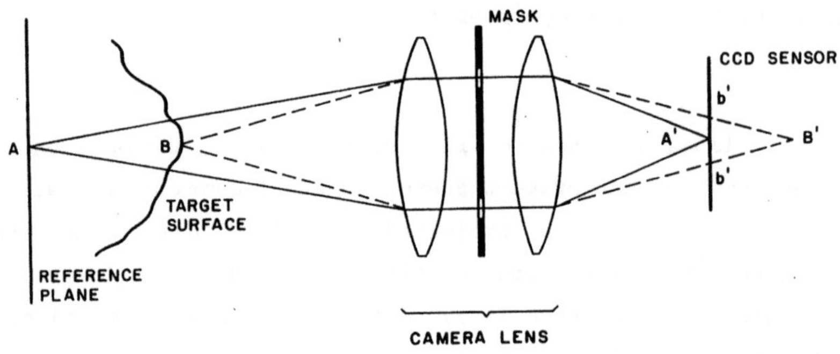

Figure 10. Triangulation using a mask in a camera lens.

compact, robust, and lightweight 3-D range sensor.[8] As seen in Figure 10, because of the two apertures in the mask, a projected point B on the object imaged out of focus at B' behind the photodetector will create two points b' on the photodetector. The separation b between these two points b' is related to the distance z by

$$z = (\ell^{-1} + Kb)^{-1} \tag{13}$$

where

$$K = (\ell - f)/fd\ell \tag{14}$$

and ℓ is the distance between the reference plane and the camera. As in all triangulation-based sensors it can be shown that the gain M_{3D} of the camera is also given by equation 8, with $\Delta p = \Delta b$.

By projecting a set of parallel lines perpendicular to the axis formed by the two apertures, pairs of lines will be measured on the photodetector, the separation between the lines being related to the height of the object.

The same advantages and disadvantages discussed previously, in the case of the projection of line or lines, apply here: low cost, simplicity, compactness, simultaneous acquisition of multiple profiles, but limited accuracy vs. the field of view. But the major advantage of this technique is that, as the triangulation involves only the camera head (lens) and not the projector, the camera is very robust and insensitive to shock and vibrations.

4.5 LASER PROJECTION FOR 3-D IMAGING

Using a laser for 3-D imaging provides a number of unique advantages. The brightness of the source ensures a good signal-to-noise ratio in most applications. If needed, the ambient light can be filtered out without any significant reduction of the laser light itself. This is because the laser light is emitted over a very narrow bandwidth. The most significant advantage is that most laser sources emit in a low order spatial mode. This property gives maximum light power and minimum divergence. This is the basic advantage of lasers for 3-D imaging. It means that all the power available at the projection remains in focus for an extended length along the range axis. Note that this is not the case for conventional light sources where, for an extended depth of focus along the range axis, the aperture of the incoherent light projection system must be reduced, which considerably lowers the amount of light available.

An analysis of laser beam propagation follows. Typical laser sources are considered, i.e., the He-Ne emitting at 633 nm, the Yag at 1.06 µm and the He-Cd at 442 nm. Most sources used have a wavelength within that range.

Parameters at the projection can be computed using the Gaussian beam propagation formula which is adequate for most laser sources:

$$w(z) = w_0\left(1 + (\lambda z/\pi w_0^{\,2})^2\right)^{1/2} \qquad (15)$$

where $w(z)$ is the radius of the beam at a distance z from the beam waist location, w_0 is the beam waist radius defined at e^{-2} of the maximum intensity , and λ is the wavelength of the source. As illustrated in Figure 11, typically a depth of field of the order of the field of view is required. In this analysis, the object is assumed to be in a cube of side a. Then, defining the depth of field following the Rayleigh criteria gives:

$$a = z_r = 2\pi w_0^2/\lambda \tag{16}$$

Such a constraint imposed by the diffraction limit leads to the relationship:

$$w_0 = (\lambda a/2\pi)^{1/2} \tag{17}$$

which gives the pixel (voxel) size assuming a sampling of the order of w_0. As an example Figure 12 shows the pixel size in term of the side of the cube (a) for various laser wavelengths. Alternatively, we can compute the image format as a function of the volume of view, which is:

$$I_F = a/w_0 = 2\pi w_0/\lambda = 2/\theta \tag{18}$$

where θ is the divergence of the laser beam. Figure 13 shows typical values for He-Cd, He-Ne, and Yag lasers. The image format means that for a cube of, say, 50 cm on a side, more than 2000 resolved spots are available for each axis x, y, z, using visible lasers. Referring to Figure 12, we see that it allows resolution of any cubic element having 200 μm on a side. It is clear from these numbers that a conventional system of 2-D projection is unable to reach such a depth of view at this resolution. As an example, a good photographic lens opened at 2 cm, has a depth of view in the order of 1 cm if the scene is 1 m away.

To take advantage of the extended depth of view at the projection the sensing element at the detection side must be carefully aligned. A

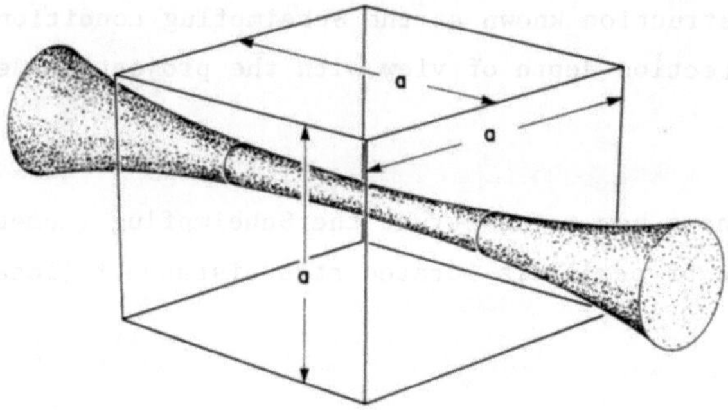

Figure 11. Gaussian beam propagation within the volume of view.

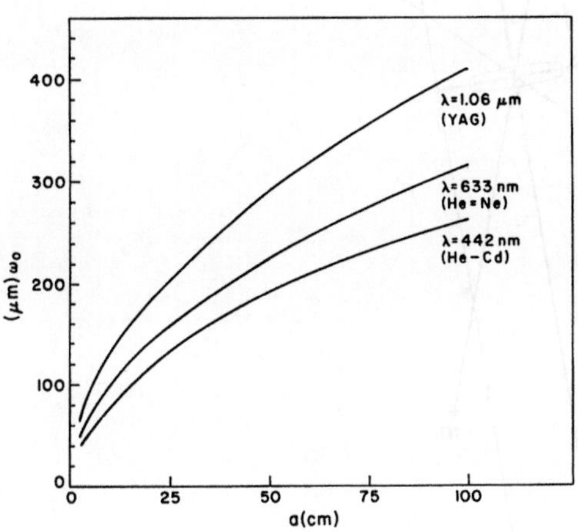

Figure 12. Diffraction limited pixels using typical laser sources.

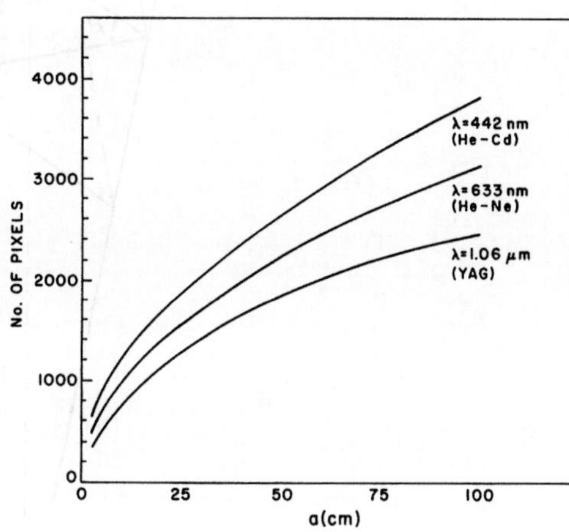

Figure 13. Image format is derived from diffraction limited pixels.

geometrical construction known as the Scheimpflug condition can be used to match the detection depth of view with the projection depth of view.

Figure 14 shows how to construct the Scheimpflug geometry. It is based on the use of parallels located at a distance f (focal length of

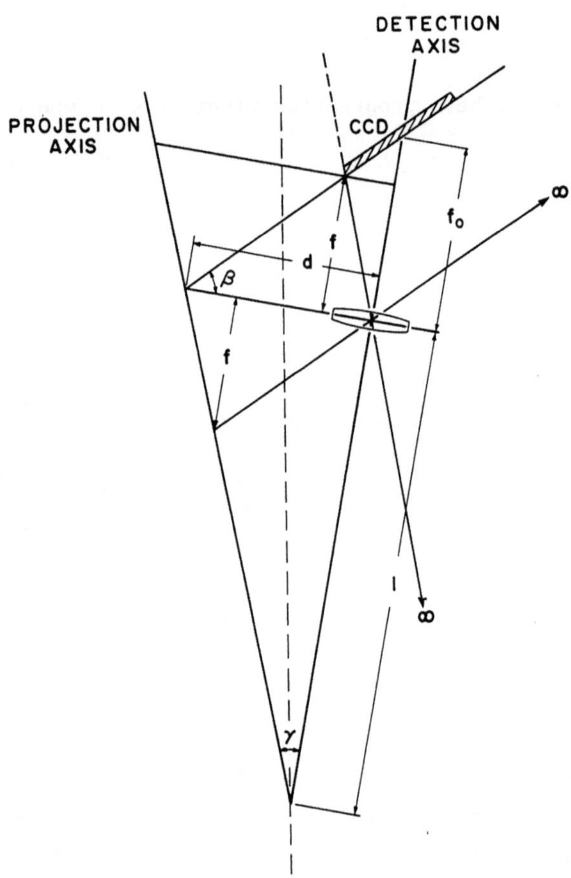

Figure 14. Scheimpflug geometry.

the lens) from the principal plane of the lens. From this arrangement:

$$\beta = \tan^{-1} (f_0/d) \tag{19}$$

This geometry is interesting because any point along the projection axis is in focus on the position sensor. This property provides a considerable improvement in the depth of view without compromising the collected energy (the lens aperture can be kept at maximum).

For design purposes, it is also convenient to estimate the image size of the laser spot scattered by the object surface. The magnification of the imaging system is easily computed using the usual definition for lateral magnification;

$$M_\ell = f_0/(\ell \cos \beta) = f/\left((\ell - f)(\cos \beta)\right) \tag{20}$$

This ratio is useful for matching the spot size of the laser beam on the object with its desired image size on the position sensor. With a sub-pixel resolution peak detector[9], it is important to have an image of the beam which is much wider than a pixel. Typically the magnification factor is between 0.1 and 1.0.

Longitudinal magnification or range magnification can be defined as the ratio of the displacement (Δp) of the laser spot image on the position sensor to the change in range (Δz) in the scene. Using Figure 15:

$$\begin{aligned} M_{3D} &= \Delta p/\Delta z = (f_0 \sin \gamma)/(\ell \cos \beta) \\ &= M_\ell \sin \gamma \end{aligned} \tag{21}$$

This relationship shows a very strong variation with distance. As an illustration, Figure 16 shows the CCD cell projected back into the

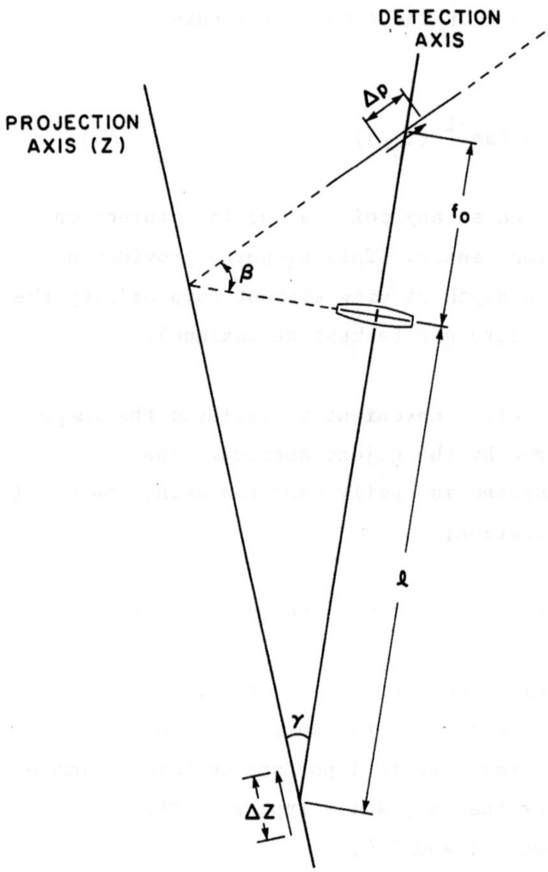

Figure 15. Range magnification on the detection axis.

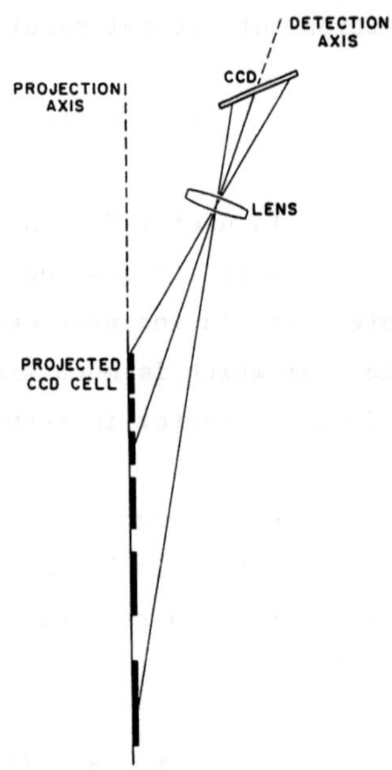

Figure 16. CCD cells are imaged on the projection axis when the Scheimpflug condition is used.

projection axis. For distant objects, the projected cell size increases due to lateral magnification M_ℓ and also to a decrease in the angle between the projection axis and the line connecting the lens with the object point of interest.

4.6 SYNCHRONIZED LASER SCANNERS

At NRC laboratories prototypes of 3-D laser camera have been built and tested.[10, 11] They are imaging devices based on synchronized scanners.

The analysis starts with the usual triangulation geometry. In Figure 17, a beam of light originates from position d along the x axis. The light beam projecting at an angle θ_0 defines a reference point $(d/2, \ell)$ that will be used for calibration purposes. At the origin there is a lens (not shown) of focal length f used to focus the light on a position sensor aligned parallel to the x axis and in focus at $-f\ell/(\ell - f)$ along the z axis. It is assumed here that d, ℓ, and f are known and are, respectively, the distance between the scanner axis of rotation and the principal point of the lens, the distance between the common axis of projection and detection (0, 0 - d, 0) and the reference point, and the focal length of the lens used to collect the light from the scene. Under rotation of the scanner, the light beam rotates to another angular position $(\theta_0 + \theta)$ (θ is negative in Figure 17). The spot of light on the position sensor moves from $df/2(\ell - f)$ to a new location p due to the intersection of the projected light beam with the object surface at x, z. The relation between the coordinates (x, z) and the parameters of the geometry is

$$x = dp\left[p + \frac{f\ell(2\ell \tan\theta + d)}{(\ell - f)(d \tan\theta - 2\ell)}\right]^{-1} \qquad (22)$$

$$z = -d\left[\frac{p(\ell - f)}{f\ell} + \frac{2\ell \tan\theta + d}{d \tan\theta - 2\ell}\right]^{-1} \qquad (23)$$

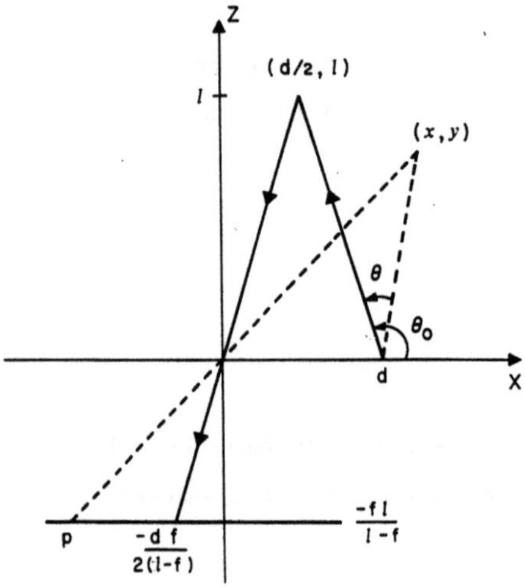

Figure 17. Laser scanner in a conventional triangulation geometry.

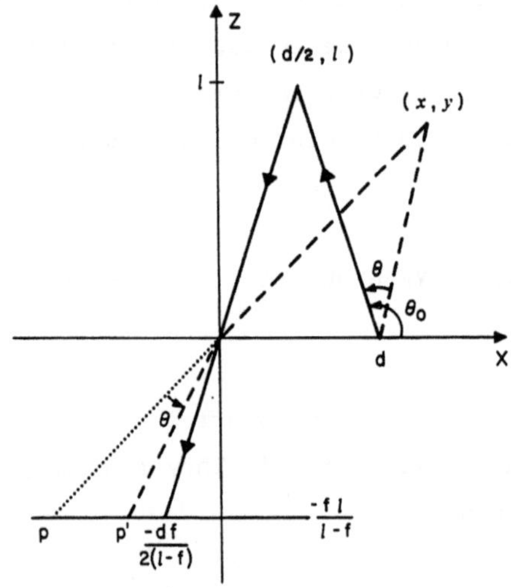

Figure 18. Laser scanner in a synchronized scanning geometry

Consider now a synchronized scanning geometry as shown in Figure 18. The only difference here is the addition of a scanner which has its axis of rotation set at (0, 0). The scanner moves synchronously with the scanner at point (d, 0) (the projection scanner) and has the effect of cancelling its angular movement. The net result is to bring the position of the spot on the sensor closer to the reference point $-df/2(\ell - f)$. That is illustrated in Figure 18 as a trajectory from point x, z to point p' and point p. Position p is the position that would be measured without synchronization. Therefore, to find coordinates (x, z) with synchronization, relate p to p', which is

$$p = \left[\frac{p' + f\,\ell\,\tan\theta}{\ell - f}\right]\left[1 - \frac{p'(\ell - f)\tan\theta}{f\ell}\right]^{-1} \qquad (24)$$

and use equations (22) and (23). It can be seen from Figure 18 that a change in the position of the light spot along the z axis produces an equivalent angular shift for both geometries (usual and synchronized), but a change along the x axis produces a much smaller angular shift in the case of the synchronized geometry. In other words, the position sensor is mainly used to measure range with synchronization, while in commonly-used geometries a large portion of the position sensor area is also used to measure the x coordinates. That is interesting because the scanner mirror position can be precisely obtained. The net gain with sychronization is that, with the same position sensor used for usual geometries, the focal length of the lens used to collect the light can be increased and an increased resolution in range obtained without reduction of the field of view.

5. APPLICATIONS

This Section discusses the applications of synchronized laser scanners developed at NRC laboratories. The first group of applications is in the area of measurement and inspection. The requirement is to obtain an accurate measure of the shape of the surface to see how it conforms to a reference. The surface is viewed from a known direction so there is little uncertainty about what is being measured. The following applications falling into this category will be described: marine propeller blade measurement, measurement of museum artifacts, entering 3-D object data into CAD systems, reproduction of 3-D object, and inspection of printed circuit boards. These are representative of the range of object sizes, resolutions, and surface properties that can be readily accommodated by this technology.

5.1 MARINE PROPELLER BLADE MEASUREMENT

This application[12] was developed under licence by Hymarc Engineering Ltd. for the Department of National Defence. Current methods for measuring large marine propellers (4 m to 8 m in diameter) rely on a variety of machines that measure a series of selected points on the face and back of the blade to determine section profile, thickness, and pitch. For a 4 m diameter propeller, 80 to 100 points would normally be measured on the face of the blade with a grid spacing in the order of 200 mm, assuming a blade width of 2 m. Depending upon the equipment, such measurements would take from several hours to a full day.

The camera designed jointly by NRC and Hymarc for this application, based on the synchronized scanning principle, has a field of view of 60 cm, a depth of field of 50 cm, and an accuracy of 0.25 mm, and can acquire 1000 points/s. The camera head measures 26 cm × 19 cm × 9 cm, has a single axis optical scan and is mounted on a 5-axis robot. It can measure on a face of a blade 5000 points on a grid spacing of 4 mm in 10 min.

5.2. MEASUREMENT OF MUSEUM ARTIFACTS

A relatively fast, noncontact surface coordinate data acquisition system has a number of potential applications in the museum field:[13]

- basic measurement of three-dimensional artifacts and specimens for documentation and research purposes.
- assessment of the accuracy of replicas.
- comparison of the shape of an object at different periods of time, for example before and after conservation treatment, before and after a loan for exhibition, or to investigate weathering, deterioration, or dimensional change over time.

- generation of digital data and contour plots which can be used either with manual or numerically controlled (NC) milling machines to fabricate replicas (at any scale), display mannequins, or supports for objects of complex shape.

To demonstrate the feasibility, a number of objects were measured including a totem pole, a preserved toad, a corroded cross, bones, etc.

5.3. ENTERING 3-D OBJECT DATA INTO CAD SYSTEMS

Creation of geometric CAD databases is usually one of the most complex parts of computer-based design and analysis systems. For objects with geometric regularity, it is practical to generate them analytically, using one of several geometric modelling schemes. However, there are many areas in which there is a need to create a database by extracting an object definition from a complex real life object, i.e., one that already exists and which does not have regular geometric properties. Examples of such objects are found in the field of medicine (prosthesis, plastic surgery, etc.), in industrial applications, such as measurements of castings, as well as in computer simulation, in which the definition of a mesh for a finite element program can be acquired automatically.

5.4. REPRODUCTION OF 3-D OBJECTS[14]

The acquisition of 3-D data of objects opens up the following possiblities:

- Producing a copy of the object.
- Producing a modified version of the object.
- Producing a negative of the object as, for example, to produce a mold.

For such applications filtering algorithms have been developed to remove the acquisition noise, which, even if minimal, must be reduced. Also in the large data arrays (256 × 256 and more) produced by the 3-D sensor, usually far more points are obtained than are needed to represent a typical object. In order to reduce the processing time of the next steps, an algorithm was developed to perform intelligent resampling of the points, based on a cubic B-spline technique. This yields a data compression of over 50%. Then the data can be sent to an interactive graphic editing system which allows visualization of the 3-D data and modification of specific points or areas, for example areas where shadow effects have prevented the acquisition of necessary data points, or surfaces of simple geometrical shape which must be defined to a high tolerance.

After a satisfactory object representation has been achieved in the editor, the resulting data are passed to a polyhedral machining program that produces the tool path for a multi-axis milling machine.

5.5. INSPECTION OF PRINTED CIRCUIT BOARDS

The inspection of populated printed circuit boards presents a difficult challenge to machine vison. The speed, resolution, and reliability required are extremely high. A minimum of 25 cm^2/s must be inspected, with resolutions in the order of 50 μm, and the error rate (missed defects or false alarms) must be less than one in a thousand. Also the system must be easily and quickly reprogrammable to accommodate changes in the production, or different boards inspected on the same production line.

Intensity (2-D) based systems have all faced serious difficulties because they must rely on shadows to locate components and the lighting required to produce appropriate shadows varies with the position on the PC board due to neighbourhood components and their surface reflectance. This forces a very difficult compromise in the lighting which has to be optimized for each board.

Three-dimensional-based techniques have a definite advantage in measuring directly the depth information for which component location can be directly obtained without ambiguity. In addition the use of synchronized scanning allows a significant reduction in triangulation angle which is important when components are closely spaced.

A system has been designed and a prototype built that has lateral resolutions of 50 μm and a data acquisition rate of 200 kHz initially, to be increased to 1 MHz. It is intended for inspection of printed circuit boards with surface mounted components.

The distinguishing characteristic of the second group of application is that the identity of the object being measured is unknown and must be determined from the measurements. The prime example is in object recognition for robot vision. The output of the range camera provides distance measurements to a matrix of closely spaced grid points projected onto the visible surfaces in its field of view. Using this distribution of 3-D samples, one attempts to hypothesize on the possible orientation of known objects that would generate the observed set of surface points. The interpretation process is supported by an object model database which provides an appropriate reference for each object in the set of objects to be recognized. Reduction of the hypothesis into a reliable identification involves an extensive search and matching process within the model knowledge base.

This process is very much a two-stage problem: one of extracting 3-D shape representations from the sensor data and the second of interpreting this information for purposes of identification. The state of the art regarding progress in 3-D object recognition can be found in a survey paper by Besl and Jain.[15] From this, it is readily apparent that the quality of solutions possible at the interpretation stage is fully dependent on the reliability of the 3-D source data on which they are based. In this respect, the range camera is capable of delivering relatively good quality data on real surface shape, thus providing a very substantial advantage over competing sensing methods, many of which rely on inferring 3-D shape from 2-D measurements.

The total effort currently devoted worldwide to 3-D vision is extensive. In spite of this, practical robot vision, even when based on direct 3-D measurement, presently requires that constraints be rigidly met. This is so because techniques have not yet been developed that adequately integrate the model representation with the search and match procedures for dealing with complex object shapes in arbitrary combinations and views. Solutions have been demonstrated for recognizing real industrial parts from a limited set when viewed in isolation and in controlled positions[16] and for simple polygonal (planar surfaced) objects in arbitrary, partially occluded views.[17] However, general purpose vision that can identify complex-shaped parts in arbitrary views is still a laboratory research problem.

Adaptations of the laser scanner for wrist mounting forms the basis for a related set of vision applications in robot guidance: robot adaptive arc welding, and part acquisition.

5.6. ROBOT ADAPTIVE ARC WELDING

The design of a compact, small field of view, high accuracy 3-D laser range finder based on synchronized scanning has been completed and transferred to Servo-Robot Inc., which has realized a complete robot welding system.[18] It includes the 3-D sensor and a camera controller which transfers commands and data to and from the camera, extracts the profiles after filtering, and does template matching. This information about the seam is fed to the process control processor which controls the correction to the trajectory of the welding robot and adjusts the welding parameters.

The major advantage of this sensor for this application is its small size and low weight, and a very high immunity to the light produced by the arc due to the very small instantaneous field of view of the detector combined with the use of laser light and a narrow bandpass filter to eliminate most of the light from the arc. This sensor (and the complete system) is commercially available and has the following performance: field of view and depth of field; 60 mm, accuracy; 0.06 mm transverse, and 0.05 mm in depth; scanning frequency: adjustable up to 28 scans/s, dimensions of 115 mm × 144 mm × 55 mm and weight of 1 kg.

5.7. PARTS ACQUISITION

Another version of the wrist mounted design is being used in the research laboratory to support experiments in parts acquisition. The optical configuration is described well in reference.[11] It uses a remote He-Ne laser with a fiber optic cable for the light source and a 256 element CCD array for range detection. When carried on the robot wrist, the range profile provides information on the position of an object relative to the robot wrist, allowing interaction with parts in a controlled way. The sensor has a working range of about 10 cm (grasping distance) out to about 1 m.

Initial use of this profile scanner has been in developing "hold-site"-based algorithms for applications in binpicking where no a priori description of the objects is available.[19] The robot first makes a coarse scan of the pile of parts to determine the nearest one and the general orientation of its major axis. The next step is to collect a sequence of profiles across the major axis and by combining them to find a holdsite which is acceptable geometrically using spatial planning methods.[20]

Overall system control is provided by a multiprocessing system controller running under Harmony, a realtime multiprocessing, multitasking operating system.[21] The Harmony controller collects data from the scanner at repetition rates of 25 Hz and moves the PUMA robot via the "Alter" mode of VAL II. A LISP machine performs the higher level spatial planning and communicates with the Harmony system when it requires data or wants to move the robot. The complete system has been tested on a variety of parts from screws to rings and works well. It takes on the order of a minute per part but this can be speeded up considerably by transferring the algorithms from the LISP machine to the Harmony system.

For this application, the robot itself was used to calibrate the profile scanner, limiting measurement accuracy to the order of 1 mm. The resolution of the sensor data is considerably better. A calibration facility is just being completed which will calibrate the instrument to accuracies of about 0.1 mm at close-in distances near the grasping range.

6. CONCLUSION

A number of techniques for 3-D imaging were reviewed. Special emphasis is placed on the triangulation theory and on the use of coherence properties of laser beams in order to get the desired depth of view. The latter is shown to be essential for close range imaging. Synchronized laser scanners are also reviewed along with their applications.

BIBLIOGRAPHY

1. M.D. Levine, et al. "Computer determination of depth maps", Computer Graphics and Image processing 2, p. 131-150, 1987.

2. A. Gerhard, et al. "Depth extraction by stereo triples and a fast correspondence estimation algorithm", Proc. Int'l. Conf. Pattern Recognition, p. 512-515, Paris 1986.

3. H. Baker, and R. Bolles, "Epipolar-plane image analysis: a technique for analyzing motion sequences," In Proc. Workshop on Computer Vision, Representation, and Control, p. 168-178, Bellaire MI, 1988.

4. R.A. Lewis, and A.R. Johnston, "A scanning laser rangefinder for a robotic vehicle", In Proc. Int. Joint Conf. Artif. Intel., p. 762-768, Cambridge, 1977.

5. D. Nitzan, et al., "The measurement and use of registered reflectance and range data in scene analysis", Proc. IEEE, 65, p. 206-220, 1977.

6. V. Srinivasan, and M. Halioua, "Automated phase measuring profilometry; a phase-mapping approach", Appl. Opt. 24(2), p. 185-188, 1985.

7. P. Vuylsteke, and A. Oosterlinck, "3-D Perception with a single binary coded illumination pattern", Proc. SPIE, 728, p. 195-202, 1986.

8. F. Blais, and M. Rioux, "Biris: a simple 3-D sensor", Proc. SPIE, 728, p. 235-242, 1986.

9. F. Blais, and M. Rioux, "Real-time numerical peak detector", Signal Process. 11, p. 145-155, 1986.

10. M. Rioux, "Laser range finder based on synchronized scanners", Appl. Opt., 23(21), p. 3387-3844, 1984.

11. M. Rioux, et al., "Design of a large depth of view three-dimensional camera for robot vision", Optical Engineering 26(12), p. 1245-1250, 1987.

12. F.R. Livingstone, and M. Rioux, "Development of a large field of view 3-D vision system", Proc. SPIE, 665, p. 188, 1986.

13. J.M. Taylor, et al., "Applications of a laser scanner to the recording and replication of museum objects", Proc. of ICOM Committee for Conservation 8th Triennial Meeting, Sydney, Australia, Sept. 1987.

14. P. Boulanger, et al., "Interface between a 3-D laser scanner and a CAD/CAM system", Proc. of the 5th Can. CAD/CAM and Robotics Conf. and Exp., p. 731-737, Toronto, 1986.

15. P.J. Besl, and R.C. Jain, "Three-dimensional object recognition", Comput. Surv., 17(1), p. 75-145 1985.

16. M.R. Ward, L. Rossol, and S.W. Holland, "Consight: A practical vision based guidance system", Proc. 9th International Symposium on Industrial Robots, Washington, D.C., March 1979.

17. M. Dhome, and T. Kasvand, "Polyhedra recognition by hypothesis accumulation from range data", NRC/ERB-985, May 1986.

18. B. Beranek, J.P. Boillot, and F.P. Ferrie, "Laser sensor for adaptive welding", Proc. SPIE, 665 p. 195, June 1986.

19. G. Roth, and D.H. O'Hara, "A holdsite method for parts acquisiton using a laser rangefinder mounted on a robot wrist", IEEE Int'l. Conf. on robotics and Automation, Raleigh, N.C., March 1987.

20. G. Roth, "Determining grasp position for a parallel type robot gripper", NRC/ERB-984, January 1986.

21. W.M., Gentleman, "Using the Harmony Operating System", Revised March 1987. NRC/ERB-966. NRC 27469

AN EXPANDABLE MULTIMICROPROCESSOR SYSTEM FOR VISUAL INSPECTION

A. J. Padilha, A. J. Campilho

DEEC - Faculty of Engineering, University of Porto

Rua dos Bragas, 4099 Porto Codex

PORTUGAL

Abstract

This paper describes a multiple microprocessor architecture for visual inspection applications where a repetitive execution of a single algorithm is required. The configuration of an application system is readily achieved by means of connecting hardware modules, according to a connection scheme that results from the algorithm, previously developed on a modular software system. This system uses an interactive command language that can be translated into a graph of basic image operations; the same graph describes the hardware network to be assembled. The operation of the network requires only the downloading of the codes of specific tasks to each processor; the initialization, identification and communication procedures are integral in each module.

An auxiliary neighbourhood processor may be used where required, to achieve an adequate throughput rate.

The system has been used in some case studies for demonstration purposes; a demonstrator of bottle cork grading is described.

1. Introduction

The recognition that conventional sequential processors are not suited, namely in terms of speed, for image processing and analysis problems, has led to the proposal of many computing architectures to overcome those inadequacies. Most of these architectures are capable of efficiently solving the iconic image processing operations, relying on standard general purpose computers for the final stages of analysis. This is a very reasonable approach [1], because the set of fundamental operations in iconic image processing is more

NATO ASI Series, Vol. F 64
Sensory Robotics for the Handling of Limp Materials
Edited by P. M. Taylor

easily identified than in later stages and also because the impact of these operations on overall system performance is very strong, as it corresponds to the stages where an enormous volume of data is present. Furthermore, the low level image operations lend themselves very well to parallel processing, unlike higher level operations, where a unified data format is not possible and sequential search may be required.

The proposed architectures exploit the inherent parallelism in low level image processing in one or several of the levels of parallelism present [2]; they can rely on concurrently effecting different operations on a sequence of images, as in data flow or pipeline architectures; or they can capitalize on dividing the image data space, often up to single pixels, and assigning each of many identical processors to each subimage, as in array processors. In either case, they can exploit the neighbourhood and the bits per pixel levels of parallelism.

In the specific domain of automatic visual inspection the same general principles apply, although systems are often tailored to restricted classes of applications or to restricted inspection strategies.

Many systems directly acquire good binary images, reduce image data by coding techniques, and then operate over the reduced data with a general purpose microcomputer (e.g., Autovision II [3], Consight-I [4]).

In other approaches, use is made of a set of specialized processors, the functions of which are associated with some specific inspection methodology. For example, SAM [5] uses specialized modules to assist in the extraction of geometric and topology features to be used in classification by discriminant analysis; GENESIS 2000 [6] uses special modules to efficiently execute iterative operations of mathematical morphology; PAP [7] uses hardware modules suited for a template matching strategy.

In another category of systems, a number of general purpose microprocessors is arranged in some convenient configuration (e.g, the concurrent processor [8]); in this approach, an increased flexibility is obtained, with relative losses in speed.

The system described in this paper is a member of the last category; the close relationship between this system and a development laboratory system is emphasized throughout the paper; section 2 characterizes the application domain primarily addressed, and lists the overall development phases; the hardware implementation is considered in section 3, whereas section 4 is devoted to system software; section 5 contains a discussion of system performance; a specific application is described in section 6, to demonstrate the general development and implementation issues.

2. Application Development

The industrial use of automatic visual inspection systems relates to objects where there is the need, and often the practice, of meticulous inspection to identify shapes, to measure distances, position and orientation, or to detect surface properties. The specification of inspection goals is often made at a high level of abstraction, suited for the human "sensor", which results in a wide range of possible implementation methods and offers the system designer an opportunity for experimentation, ingenuity and intuition. As a consequence, the available methods and techniques tend to proliferate.

In this context, application development is usually made in two steps: first, in a laboratory environment, methods are tested, in general purpose systems, to demonstrate and validate competence; then, in a second step, the selected solution is adapted to an existing real time system or a new dedicated system is designed for industrial use.

The multimicroprocessor architecture herein described was designed as a means to alleviate the difficulties found in the second step of the development process.

The aim has been the design of an implementation system, closely related to the laboratory development system, and allowing the very fast production of industrial prototypes with a precisely predictable performance and cost, thus obtaining a fully integrated development system.

In this section, we first define some characteristics of the application domain to be addressed, then we review the basic aspects of the laboratory system in use, and, finally, we point out some guidelines and requirements for the design of the implementation system.

2.1 Application Domain

A considerable number of reported applications of visual inspection in industry occur in sectors (e.g. electronics, automotive) with a strong economic and technical structure; other industrial sectors, lacking those capabilities, may not develop and incorporate this new technology, no matter how appropriate it may be.

Recognizing this fact, and taking advantage of working in an academic environment, we intended to contribute to the dissemination of industrial visual inspection, by addressing those application domains where the main limitations are of economic nature, rather than of technical difficulty.

The introduction of visual inspection systems should not require massive investment, and it should not demand the production of a large number of units to be viable. We found that this was possible, if three critical factors could be met:

a) the ability to demonstrate, in a laboratory environment, the feasibility of the application; this can be done, in many cases, by using a general purpose image analysis system;

b) the means to design an application system capable of real time operation, that is, meeting execution speed requirements;

c) the capability to assemble and deliver the application system in a very short time; that is, achieving development speed.

2.2 Laboratory Development System

We have developed an interactive image processing system, named SPAI [9], based on a standard microcomputer (LSI-11) and including a real time frame grabber and a frame buffer; image capture is made by a standard TV camera. The image processing and analysis functions are all performed by software.

SPAI design has been inspired by SUSIE [10], retaining the concept of organizing image functions as collections of trivial image operations; a command language was defined, each basic command corresponding to one primitive operation.

Basic commands may be used in sequences and they can be combined in macro-commands or in command programs.

As with SUSIE, SPAI directly operates on two images; the *current* image which is the one displayed on the TV monitor, and the *alternate* image; the two images are exchanged as commands are executed, so that the result of an operation is always found in the *current* image.

SPAI has about 70 basic commands, in the following classes:

- *utility*, namely for image transfer to and from disk, for macrocommand creation, for lookup table definition, etc.;
- *acquisition and display*, for the obvious tasks of freezing a TV image, displaying an image on a TV monitor, or printing it on a microplotter;
- *processing*, using pixel operations over one image (e.g., thresholding, exponentiation), or two images (e.g., anding, subtracting), or using local operations in a 3 x 3 neighbourhood (e.g., local maximum, arithmetic convolution), and global operations (e.g., picture shift, zooming);

- *analysis*, to obtain relevant data structures (e.g., histograms, projections) and features (e.g., perimeter, Euler number).

A proper choice of these primitive operations offers an experienced user the ability to combine them into powerful higher level functions.

Another software system, named SARF [11], has been developed, running on the same host, to assist the user in feature extraction and selection, and in decision logic design. SARF is a menu driven software package, offering the user a set of pattern analysis and recognition utilities, grouped in several modules.

These modules have a common working buffer, to keep general information about data structure, *a priori* and *a posteriori* classifications, and data itself. Data is organized as a number of structured or unstructured feature vectors, usually measured by SPAI or by a special data analysis program. Several operations may be performed upon the data set, namely: *feature selection* using different class separability criteria; *supervised classification* with different possibilities of decision logic design (e.g., K-nearest neighbour, Bayes classifiers); *unsupervised classification* for grouping the data (e.g., leader algorithm, various versions of K-means); *feature extraction and mapping* to reduce dimensionality of data, and to create useful bidimensional graphic display of multi-dimensional data (e.g., Fisher-declustering criterion, Sammon algorithm).

At present, the LSI-11 microcomputer is replaced by a PDP-11-23+, and a more powerful commercial image acquisition, processing and display board set (including 3 frame buffers, one ALU and one histogram processor) added.

A new software system (SPIRF) has been developed [12]; it retains the main characteristics of SPAI and SARF, with increased facilities in program editing, new language constructs and a more convenient user interface.

2.3 Guidelines to Implementation System

In designing the implementation system, several specifications were considered, namely: a) fast configuration and installation of an application system; b) easy maintenance and repair ; c) reuse of system components, either in new applications or in new versions of the same application; d) possibility for autonomous operation; e) use of standard, established technology.

In terms of operational specifications, the main objectives have been: a) variable spatial resolution, while considering that the typical number of pixels across rows and columns would be in the range of 64 to 256; b) grey-level processing, allowing up to 256 grey-levels; c) range of pixel, local and global operations identical to the ones found in SPAI.

The implementation system, named AMIVA, uses some of the concepts developed in the context of special architectures for image processing:

- the image data space can be divided, in order to reduce the amount of data processed, by each microprocessor;
- the full processing and analysis algorithm can be decomposed in sequences of primitive operations, groups of which may be assigned to different processors, operating as a data flow chain; this is the key issue in AMIVA;
- the topology of connections among processors can be changed to match the operational requirements; however, the reconfiguration is merely static;
- the image memory is shared among processors, facilitating its communication; this is not a global property in AMIVA, as memory is only shared among "neighbour" processors.

3. AMIVA Hardware Implementation

In SPAI, the solution of an inspection task is expressed as a combination of basic commands; the SPAI language expression can be easily translated into a directed graph, where the nodes represent the elementary software modules (the basic commands) and the arcs convey image data from one node to another. Each node can have inputs from one or two nodes and its output can be directed to one or several nodes.

As an example, consider the SPAI language sequence:

TH R1; EB; HX; CS 0; XY; LB; EX; EB;EX; XY; HX; CS 0; LB; EE; YX

which determines the rectangular area that surrounds the object detected by TH R1 (thresholding). The corresponding graph is represented in figure 1.

The approach to AMIVA design has been to replicate in hardware modules the graph structure obtained after SPAI.

The AMIVA modules are all identical, consisting of a general purpose microprocessor and associated memory; communication between modules is achieved through shared memory. Thus, to implement the previous example, AMIVA modules should be connected as in figure 2.

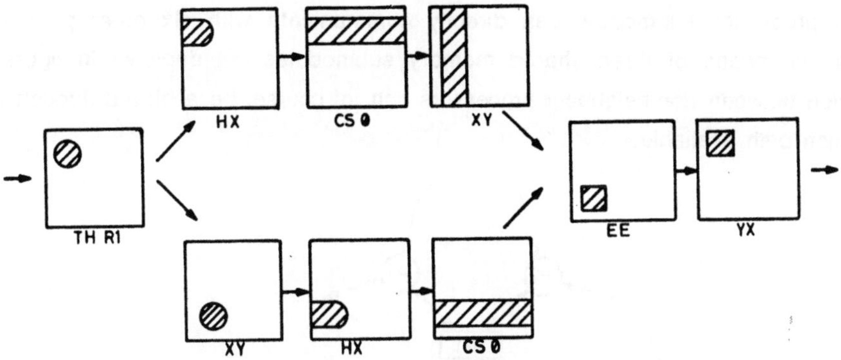

Figure 1. Surrounding rectangle SPAI graph.

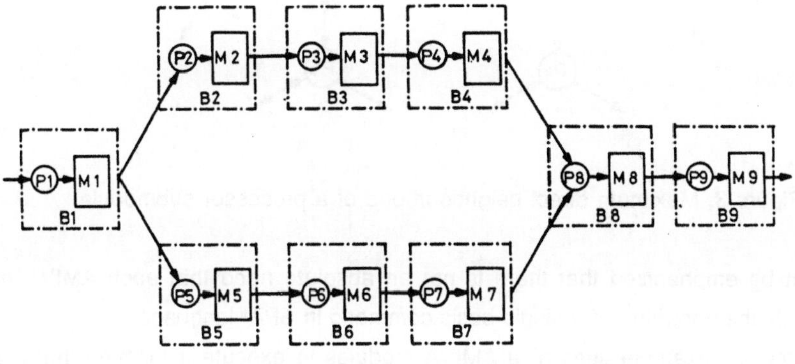

Figure 2. AMIVA modules interconnection, corresponding to the graph of figure 1.

Each module's processor must have access to, at least, two other module's memory; on the other hand, each module's memory should allow access from, at least, two other module's processors.

In fact, each module has been implemented in two separate three-ported submodules, processor and memory, as there is no reason for a processor to have "its own" memory; as a result of this separation, greater flexibility is achieved.

Each processor submodule can directly communicate with six other processor submodules, by means of three shared memory submodules, as depicted in figure 3; communication between non-neighbour processors can, of course, be achieved through any interconnection path available.

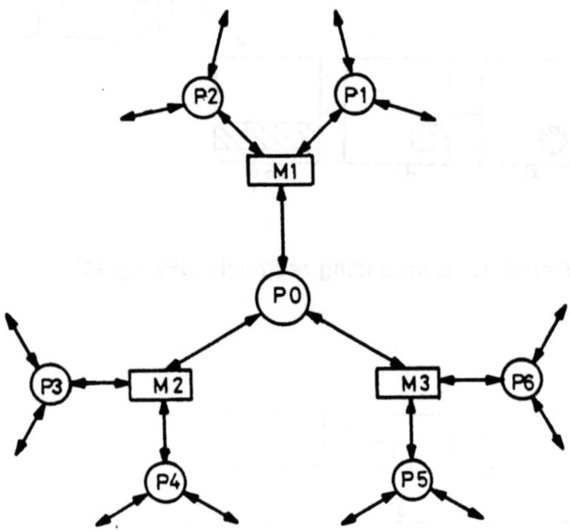

Figure 3. Maximum direct neighbourhood of a processor submodule.

It must be emphasized that there is not an absolute need that each AMIVA module performs exactly the operation of a single basic command in SPAI language.

The previous example uses nine AMIVA modules to execute a function that could be equally well performed in a single image scan, requiring a single module with a computational load equivalent to any of the nine modules in figure 2.

On the other hand, and according to the desired performance, the operation of a single SPAI module can be distributed over two or more AMIVA modules, by decomposing the function space (modules in series) or the data space (modules in parallel).

The overall operation of an AMIVA structure obeys the concept of macropipelining [13]; the operation of a module can be decomposed in two alternating macrocycles, where the memory submodule uses a double-buffering technique, so that it receives image data from one processor, while sending to another the image data it received in the previous macrocycle.

We call this type of operation *inter-image pipelining* and it requires that a memory submodule has, at least, the capacity for two images.

However, it must be noted that, very often, namely when two successive modules use the same pixel address sequencing, there is no need to buffer full images; in this case, we can buffer only small segments of image data (e.g., one row), achieving what we call *intra-image pipelining*.

Besides memory savings, the operation in intra-image pipelining mode significantly reduces the latency time from image acquisition to inspection results, with a minor effect on global throughput rate.

3.1 AMIVA Modules

As described earlier, an AMIVA module is composed of a processor submodule and a memory submodule. Assembling an AMIVA system requires the interconnection of processors to memories in an appropriate pattern; this is achieved by using a single type of connecting cable; the only connection restriction is that a cable must always link a processor submodule to a memory submodule.

To demonstrate the AMIVA concept, a number of modules was built, using existing 4MHz Z80 STDBUS boards with up to 12 Kbytes onboard RAM/EPROM. A processor submodule consists of one of these boards and a companion board (switcher), that essentially implements the three ports to access memory submodules; optionally, other auxiliary boards (e.g., for image I/O) can be placed on the STDBUS.

The switcher board implements the following functions:

- *address translation*, allowing the processor to address, through a 4 or 16 Kbyte window, 1 Mbyte of memory on each of the three ports;
- *port access control*, to select the port to activate, and to generate the corresponding access request signals;
- *memory lock function*, to reserve the addressed memory submodule for successive references, and to prevent its access from other processors;
- *wait-state control*, to force the attached processor to delay the completion of its instruction until the selected port is ready for transfer;
- *interrupt handling*, to syncronize and direct to the microprocessor the interrupts received on any port.

The memory submodule is built on a single board, if intra-image pipelining is used; a second board must be used for inter-image pipelining.

The main functions of the memory submodule are as follows:

- *memory*, which is accessible by three ports and is organized in two levels: level 1 consists of 4 Kbytes RAM and is used for control purposes and for image data in intra-image pipelining mode; level 2, implemented in the secondary board, which can be populated with up to 1 Mbyte, to serve as image buffer space in inter-image pipelining mode;
- *arbiter*, to account for possible conflicts among processors accessing memory; the arbiter generates grant signals for each port, using a round-robin discipline if more than one request is present; it is an asynchronous, independent requests arbiter, operating very fast;
- *access acknowledge*, to enable one requesting processor to resume its operation over memory; it essentially generates a delayed version of the grant signals produced by the arbiter, so that valid address and data signals are established;
- *interrupt handling*, to generate awake signals for the processors linked to the ports; two interrupt signals are generated when a processor, connected to one of the ports, performs a special dummy write cycle to memory; these interrupts are latched and sent to each of the other two ports; a dummy read cycle (part of the interrupt handling routine of the processors) clears the associated latch.

3.2 Neighbourhood Processor

AMIVA structures, operating in a pipeline fashion, can suffer from imbalance among the processing load of the various modules; it is clear that the global throughput rate is determined by the slowest module in the structure.

As mentioned earlier, the operation of one module can be distributed over two or more modules, either by decomposition in suboperations or by fragmenting the image data space; this technique can be used if one or some submodules are much slower than others, in order to achieve a faster throughput rate.

On the contrary, if some modules are much faster than others, their operations may be combined in fewer modules, so that a better balance is achieved and the module count is reduced.

There is, however, a class of operations, namely those on a 3 x 3 neighbourhood, where AMIVA modules are highly inefficient and the operation decomposition strategy would be too costly.

Having these neighbourhood operations in mind, an auxiliary processor (NP) has been designed to operate in conjunction with an ordinary AMIVA processor submodule; the NP is implemented on an STDBUS card and it communicates with the module's processor solely by I/O operations, thus performing only an accelerator function without any further effect on the AMIVA structure.

The neighbourhood operations to be performed are those found in SPAI, namely:

- arithmetic convolution of an image with a 3 x 3 coefficient window;
- 3 x 3 local logic operations on binary images;
- maximum (or minimum) in the neighbourhood;
- region labeling on binary images.

The block diagram of NP is shown in figure 4.

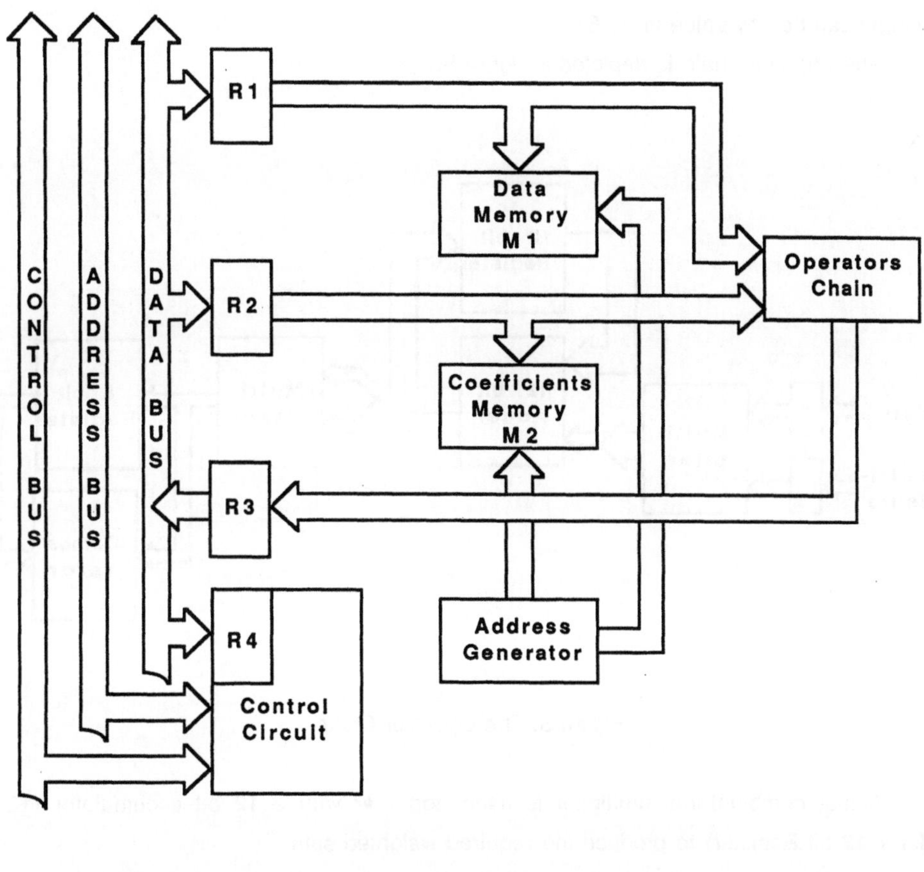

Figure 4. Block diagram of AMIVA neighbourhood processor.

The interface to STDBUS is governed by four eight bit registers; R4 is the control register and its contents determine the current mode of operation:

- in programming mode, the coefficient memory, M2, is programmed through register R2;
- in execution mode, register R1 is successively loaded with image pixels in a raster scan; loading of R1 triggers an internal processing cycle, which results in an output pixel that can be read in register R3.

NP uses an internal 10 MHz clock that, in conjunction with the address generator, sequences ten subcycles, during which the nine pixels of the current neighbourhood and their corresponding coefficients are fed to the operator chain, and the final result is normalized.

The data memory is circularly addressed, with a capacity of three image rows; the row length can be any value up to 512.

The operator chain is depicted in figure 5.

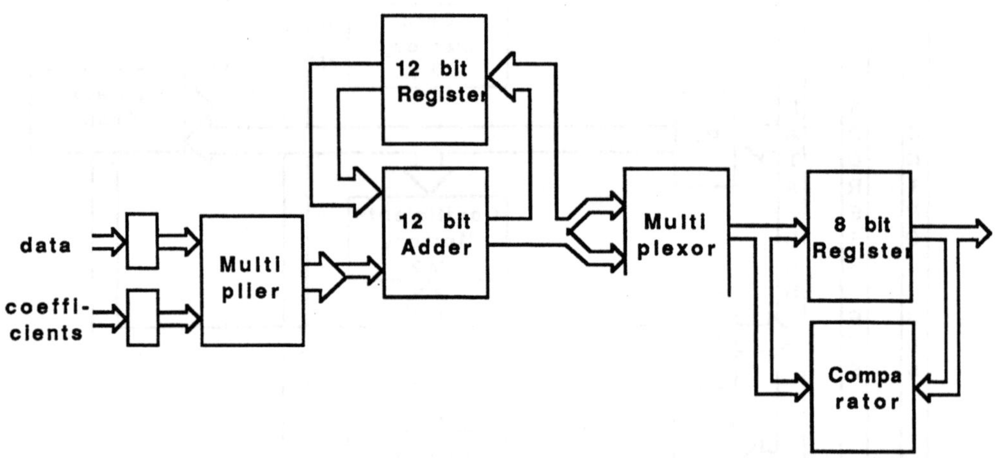

Figure 5. The Operator Chain.

A fast combinatorial multiplier is used, together with a 12 bit accumulator (12 bit Adder + 12 bit Register) to produce the required weighted sum.

In arithmetic convolution, the multiplier is programmed for mixed mode and the 12 bit Register is loaded, in the first subcycle, with a coefficient dependent constant value, so that the net result after the nine subsequent cycles is a nonnegative value; in order that the dynamic range of the result can be as large as possible, the window coefficient with the highest absolute value must be chosen close to -128, if negative, or to 127, if positive.

In local logic operations, the multiplier is programmed for unsigned mode and the window coefficients are chosen as displayed in figure 6, so that the final result is an eight bit word, each bit corresponding to the binary value of a specific pixel in the image window; the 0 coefficient corresponds to the pixel currently written into R1 (figure 4), so that the module's microprocessor can use that pixel value to select one of two lookup tables to be addressed by the value read out from register R3. Thus, any 3 x 3 logical operation can be executed.

8	16	32
4	128	64
2	1	0

Figure 6. Window coefficients for logical operations.

To configure the maximum (or minimum) in the neighbourhood operation, the multiplier and the accumulator are bypassed (in fact, they need not be present). The 8 bit Register (figure 5) is loaded with an extreme value in the first subcycle (0 for maximum; 255 for minimum) and, in subsequent cycles, this register is conditionally clocked, according to the outcome of the comparator.

Region labeling operations, over binary images, are used to assign different labels to different connected regions; all pixels in a specific connected region are assigned the same unique label. To label a pixel X, its upper and left neighbours must be considered; we shall call those neighbours A, B, C and D, according to its position, as shown in figure 7 a).

B	C	D
A	X	

a)

2	4	8
1	16	0
0	0	0

b)

Figure 7. Region labeling: a) neighbourhood configuration; b) window coefficients.

If we perform a 3 x 3 logical operation, by first changing the coefficient window to the values shown in figure 7.b), then the output value, **O**, may assume 32 values which must be interpreted in the following way:

- if $0 \leq \mathbf{O} \leq 15$ then assign label **0** to X (X is a background pixel);
- if $\mathbf{O} = 16$ then assign X a new label;
- if $\mathbf{O} > 16$ then assign X the label of
  - A, if **O** is odd;
  - B, if **O**/2 is odd;
  - C, if **O**/4 is odd;
  - D, otherwise.

The module's microprocessor must build and process an equivalence list and, to have faster access to the labels of X's neighbours, it must buffer the labels of pixels in the present row and in the preceding one.

It can be shown that the equivalence list needs to be updated only if the output value **O** is 25 (in which case the label of A is equivalent to the label of D) or if it is 26 or 27 (cases where the label of D is equivalent to the label of B); for all other values of **O**, either there is no equivalence or the detected equivalence has been listed in a previous step.

4. AMIVA System Software

It has been considered essential, in AMIVA design, that only a minor part of the operating code had to be specifically loaded in each module; the full control and communications software had to be application independent. Only the application specific image operation code should be loaded into the modules.

Moreover, it was decided that control and communications would be distributed and that every module would have a copy of a single system software.

Module processing is triggered by interrupts, after which an indefinite, interruptible waiting cycle is executed. Hardware interrupts, from one processor to another, are directly related to message passing from one process to another. A message is only sent if the receiving process is ready to accept it and, on the other hand, a receiving process waits until the sending process is ready for communication; in this way, the global AMIVA operation is self synchronized.

On power-on, or after reset, every processor submodule executes an initialization routine to detect and register the state (connected / not connected) of the three ports; then a self identification process is executed, so that each processor has an exclusive identity. After identification completion, the application specific programs are distributed throughout the structure and the system becomes ready for normal operation.

In describing AMIVA software we first examine the communication between modules and the memory organization; then, we overview the global software organization; we proceed to describe how external programs, the application specific code, must be written; and, finally, we examine the self identification and the program loading processes.

4.1 Communication and Memory Organization

The identity of processor submodules is used to specify the message sender (ORIGIN) and receiver (DESTINATION); as any processor has, at most, six neighbours, message communication requires the intervention of intermediate modules to effect message transport, when ORIGIN and DESTINATION are non neighbours; thus, every processor must have a **path table** to direct messages to any another module.

The processors that share a specific memory submodule (at most 3), assign themselves an access class (A, B or C) during initialization; the access classes are used to build the basic message passing mechanism in the memory submodule.

Level 1 memory is organized as shown in figure 8; the memory is logically divided as sixteen 256 byte areas; the first is used for communication control and the remaining 15 are used as data buffers (DEPOSITS).

The control area is further divided in eight 32 byte subareas; the first is responsible for the coordination of the access to message headers; the last subarea is used to broadcast any detected errors and the middle six are used to register the headers of the messages among the processors with access classes A, B and C.

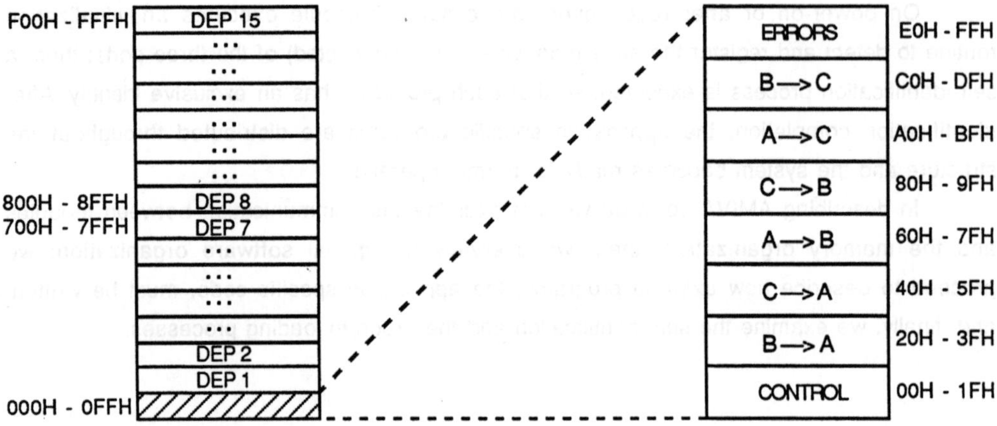

Figure 8. Level 1 memory organization.

The construction of the message headers is free, except for the first 4 bytes which must contain the sequence

CODE, DEPOSIT, ORIGIN, DESTINATION

where CODE specifies the address for a **code table** containing the execution addresses of the implemented functions (either external programs or system programs such as self identification or error processing); DEPOSIT specifies the number of the deposit where the first data are to be found (if zero, the message is header only); ORIGIN and DESTINATION contain the identities of the sender and of the receiver, respectively.

Although message handling details are beyond the scope of this paper, it must be noted that the data part of a message is not limited to the length of one deposit (256 bytes); in fact, the message data can have any length, as the communications control area keeps information on continuation data deposits.

4.2 Software Organization

It was mentioned earlier that each processor submodule has a copy of a single operating system; that copy is contained in a 2 Kbyte EPROM called SISROM.

SISROM is located in the private memory of the microprocessors, according to the map depicted in figure 9.

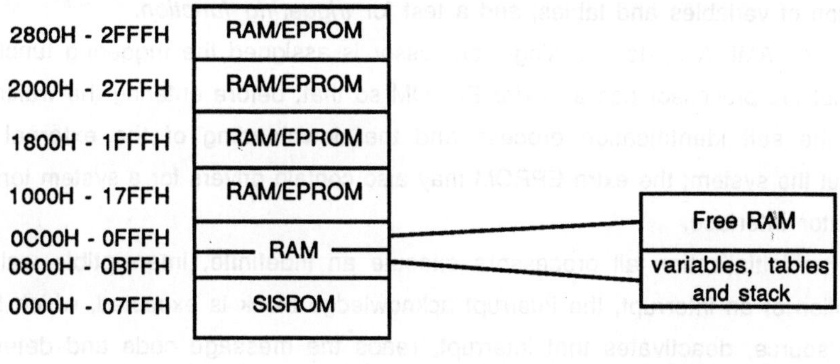

Figure 9. Microprocessor private memory organization.

SISROM and 2 Kbyte of contiguous RAM are required for system operation; the lower 1 Kbyte of RAM is used for system variables, tables and stack; the upper 1 Kbyte may be used to store the external programs. The block diagram of system operation is briefly described in figure 10.

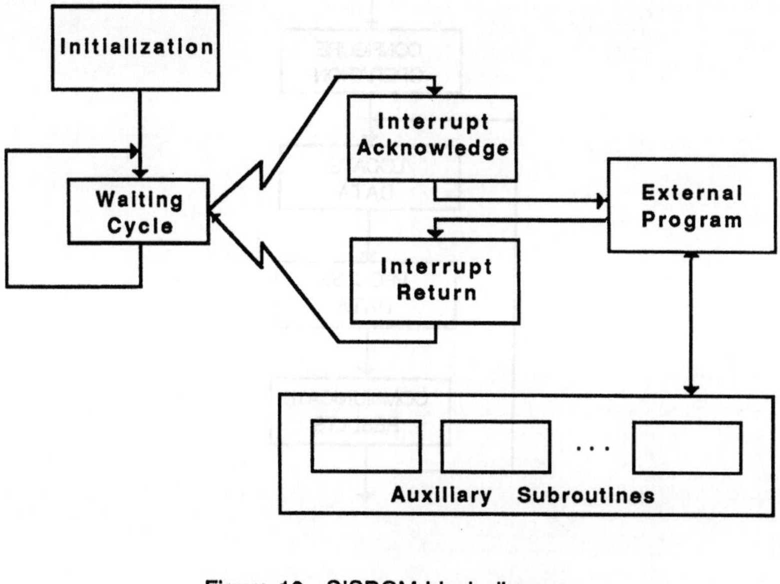

Figure 10. SISROM block diagram.

The code for all blocks is contained in SISROM, except for the external program block.

The initialization procedures, executed by all processors in AMIVA after power on, contain tests of private memory extension, access class determination for all three ports, initialization of variables and tables, and a test for *triggering function*.

In any AMIVA system, a single processor is assigned the triggering function, which means that the processor has an extra EPROM so that, before entering the waiting cycle, it initiates the self identification process and the downloading of the external programs throughout the system; the extra EPROM may also contain drivers for a system terminal used for operator interface.

After initialization all processors execute an indefinite, interruptible waiting cycle; on reception of an interrupt, the interrupt acknowledge block is executed, which locates the interrupt source, deactivates that interrupt, reads the message code and determines the corresponding execution address.

4.3 External Programs

The general block diagram of an external program is as shown in figure 11.

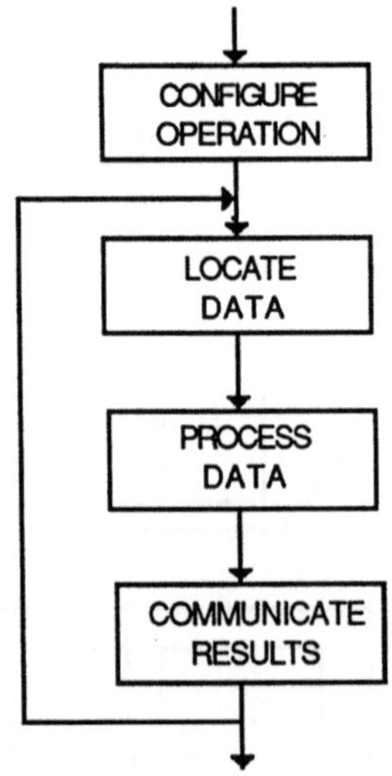

Figure 11. Block diagram of an external program.

The configure operation block essentially reads the message header and extracts any relevant parameters, such as image row length and threshold value ; the process data block refers to the operation to be performed. In fact, only these two blocks require specific code, as the others may call auxiliary subroutines resident in SISROM.

Some of the most relevant subroutines available perform such functions as:

- locating the start address of the first input data deposit and of the continuation data deposits;
- freeing an input data deposit, making it available for new data;
- locating and writing the header of a message to send;
- blocking and unblocking output data deposits before and after outputting data;
- sending a message, by activating the required interrupt signal.

4.4 Self Identification Process

A dynamic identity assignment process is used in AMIVA systems, ensuring that all processor submodules are identified and that the initiating processor (the one with the triggering function) receives full information on the interconnection structure.

The self identification process uses the basic message passing mechanism to "travel" through the entire system. The principle can be briefly stated as follows:

- a processor that receives the self identification message (a code 1 message) for the first time (the sender is the "parent" processor), retrieves its identity from the header; it then selects one of its neighbours and sends it the same message, with a new identity; it then awaits reception of the corresponding reply, containing interconnection information, as seen by that particular neighbour; the process is repeated to all "descendent" neighbours;
- when a processor receives a self identification message, and it already has its own identity, it immediately replies, showing its identity;
- when the initiating processor has received a reply from all its neighbours, the self identification process terminates, and the full interconnection scheme is available.

To illustrate the method, consider the AMIVA system shown in figure 12, where the labeled circles represent processor submodules and the small dark spots stand for memory submodules. Note that the full structure is identified for explanation purposes only. The initiating process is the one with 0 (zero) identity.

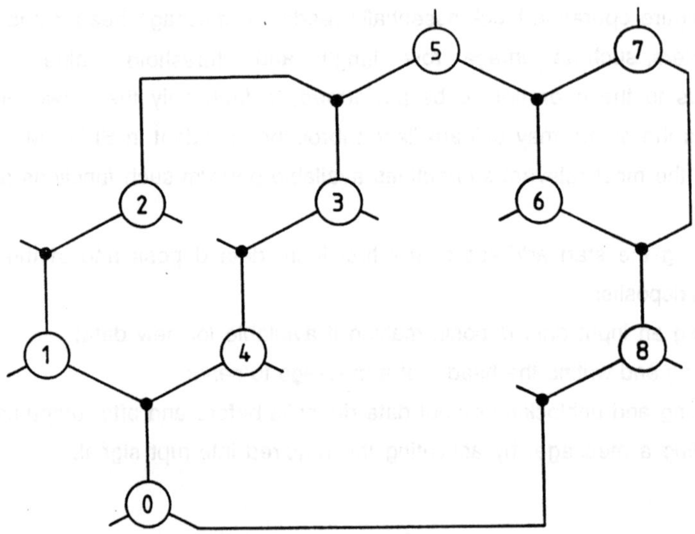

Figure 12. Identified AMIVA structure.

Figure 13 contains a compact description of the full sequence of messages, denoted by arrows, passed among processors to obtain the identification pattern of figure 12.

Messsages labeled (a) are those sent to an unidentified neighbour; those labeled (b) are those sent to an already identified neighbour, which replies with a message labeled (c); messages labeled (d) are those containing the reply informatin to the "parent" processor. Note, however, that those labels convey only a logical distinction, as all messages are of the same type and they all use code 1.

4.5 Program Loading

After self identification, the full message passing structure is established. The next step consists in loading the external programs into the modules; this is achieved by using a subroutine also contained in SISROM, that allows general data transfer between any two modules in the system; if a special byte in the message header contains a nonzero value, the destination processor interprets the received data as program code and places it at the address specified in the header.

In general, the program loading process is conducted by the processor with the triggering function, which supposedly has access, either in EPROM's or in mass storage, to all external programs required by the application.

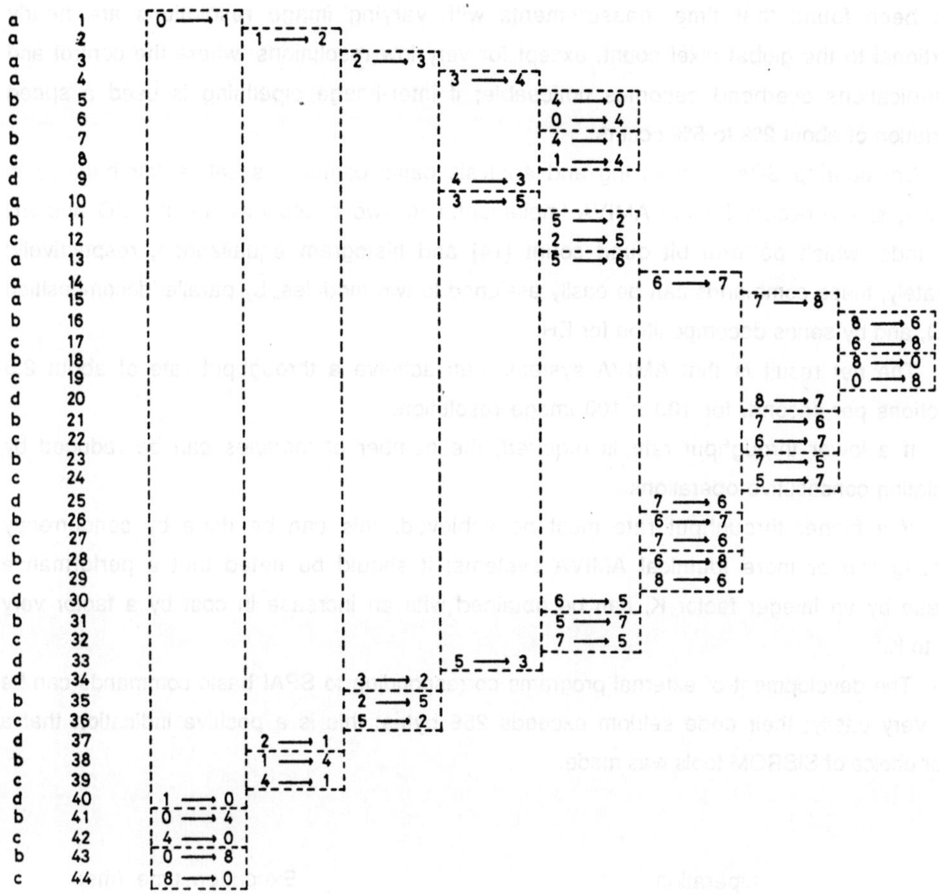

Figure 13. Full sequence of messages in a self identification process.

5. Performance Evaluation

While using a very conservative technology, the AMIVA demonstration system showed a reasonable performance, adequate to the application domain considered.

Six modules were built and tested in various interconnection patterns, executing many SPAI-like operations.

Some experimental results are listed in table 1; these results refer to the processing time of a single module, in intra-pipelining mode, for a 100 x 100 spatial image resolution.

It has been found that time measurements with varying image resolutions are nearly proportional to the global pixel count, except for very low resolutions, where the control and communications overhead becomes noticeable; if inter-image pipelining is used a speed degradation of about 2% to 5% occurs.

Considering SPAI processing and analysis basic commands set, a fair balance in operating speed occurs for the AMIVA implementation; two exceptions are the BQ and EH commands, which perform bit quad count [14] and histogram equalization, respectively; fortunately, these commands can be easily assigned to two modules, by parallel decomposition for BQ, and by series decomposition for EH.

The net result is that AMIVA systems can achieve a throughput rate of about 2.5 inspections per second, for 100 x 100 image resolution.

If a lower throughput rate is required, the number of modules can be reduced by associating consecutive operations.

If a higher throughput rate must be achieved, this can be done by concurrently operating two or more identical AMIVA systems; it should be noted that a performance increase by an integer factor K, can be obtained with an increase in cost by a factor very close to K.

The development of external programs corresponding to SPAI basic commands can be done very easily; their code seldom exceeds 256 bytes; this is a positive indication that a proper choice of SISROM tools was made.

Operation	Execution time (ms)
- Lookup table (LUT) transformation	317
- Binary thresholding	269
- LUT transformation followed by binary thresholding	391
- Grey level histogramming	232
- Binary horizontal projection	247
- Grey level image addition	378
- 3 x 3 neighbourhood operations (except region labeling)	305

Table 1. Performance measures (100 x 100 images).

6. Visual Inspection of Bottle Corks

Bottle cork grading presents several problems, mainly resulting from the biological nature of this product. In fact, we found a wide variation in streak and spot sizes, shapes, orientations, and in overall colouring and shading. As a consequence there is no defined criteria to enable the discrimination of the different (usually six) classes. This difficulty is confirmed by the low level of agreement (about 40%) between a pair of workers in classifying the same lot of bottle corks.

To overcome these problems we devised a visual inspection strategy that may be decomposed in the three following main steps: 1) preprocessing of a bottle cork grey level image, with 6 bits/pixel and a 50 X100 resolution; 2) image analysis of the preprocessed image and, 3) according to the features extracted in step 2), grading the bottle corks, into different classes. In this section we present an integrated approach to "bottle cork grading", using the laboratory facilities described in the previous sections.

6.1 Image Preprocessing and Analysis

Colouring and shading variations on the bottle cork surface as well as different levels of illumination, prevent the use of a straightforward and fast threshold operation do detect the defective region. Instead, a defect detector, insensitive to these local and global changes, must be used. A variety of image processing techniques were tested in SPAI system. The most successful one was found to be the streak and spot detector [15]. The result of this algorithm is a binary image with a map of all defective regions.

Let $f(i,j)$ and $g(i,j)$ be images with N x M resolution where $1 \leq i \leq N$, $1 \leq j \leq M$; **max**(f) and **min**(f) are the local maximum and minimum 3 x 3 operators; **sub**(f, g) is the image subtraction and **th**(f) is the thresholding of $f(i,j)$. The defect detector may be described by the following steps:

Step	Operation	
1	$g(i,j)$ <— $f(i,j)$	
2	$f(i,j)$ <— 3 * **max**(f)	(a)
3	$f(i,j)$ <— 3 * **min**(f)	(a)
4	$f(i,j)$ <— **sub**(f,g)	
5	$f(i,j)$ <— **th**(f)	

((a) - 3 * **max**(f) or 3 * **min**(f) means that the **max** and **min** operators are repeated three times)

This algorithm has a number of interesting and adequate features for bottle cork defect detection. In fact, the result (a binary image with a map of the defective region) is independent of cork colouring, shading and illumination. Also, we may tune the maximum dimension of defects to be detected, adjusting the number of cycles used in steps two and three. Figure 14 shows the result of this algorithm for two bottle corks of different quality.

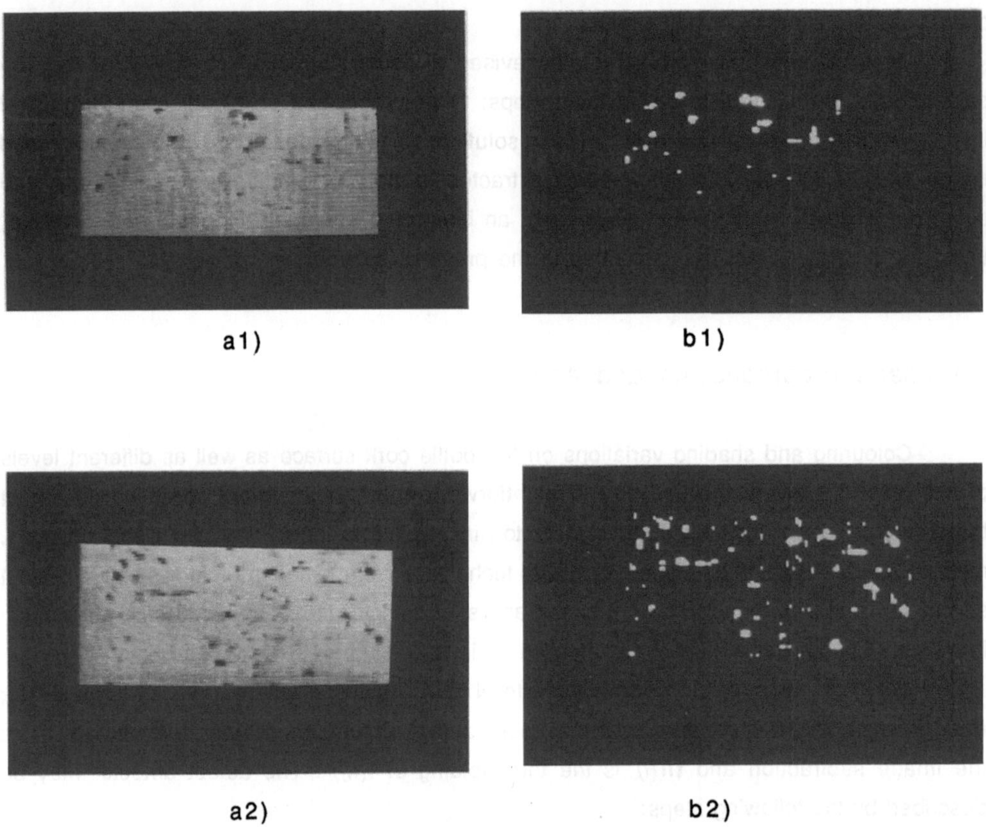

Figure 14. Defect detector: a1), a2) original images;
b1), b2) results of defect detector algorithm.

The presence of major defects negatively affects the cork quality. To obtain a map of the major defects a sequence of opening - **open***(f)* - and closing - **close***(f)* - operators were used, according to the following steps:

Step	Operation		
1	*b2(i,j)*	<—	*b1(i,j)*
2	*b2(i,j)*	<—	**open***(b2)*
3	*b2(i,j)*	<—	**close***(b2)*
4	*b2(i,j)*	<—	**and***(b1,b2)*

where *b1(i,j)* is the binary image obtained after the defect detector and the operator **and***(.)* is the logical anding of two images. The result appears in the binary image *b2(i,j)*. In figure 15 we can appreciate the result of this algorithm. For convenience the image with a map of defects will be called *def(i,j)* and the one with the major defects *maj(i,j)* .

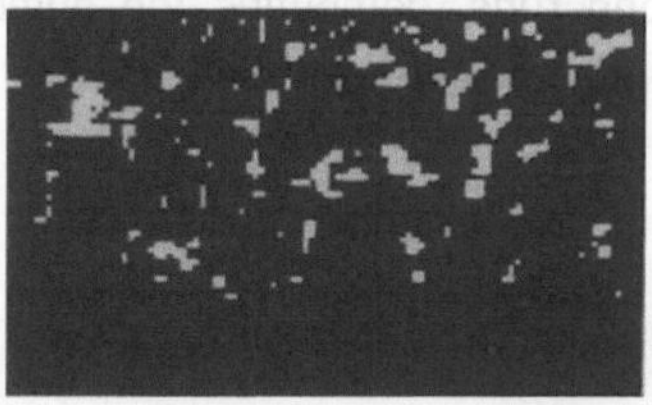

a)

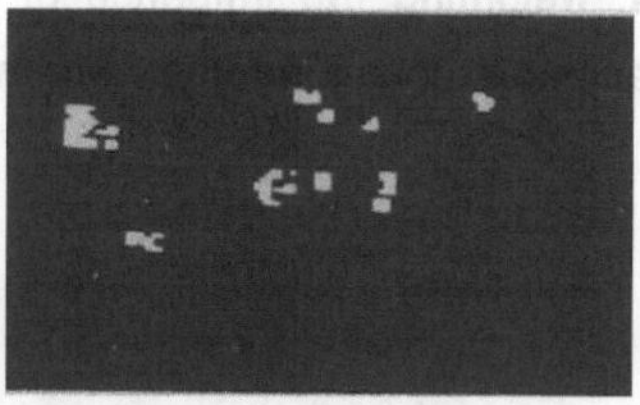

b)

Figure 15. Major defects detection: a) original image; b) image result.

The main goal of the image analysis task is to extract a meaningful set of features that characterize some well known aspects referenced by human classifiers. As we have already mentioned, the main difficulty presented by this task is the absence of objective criteria that enables the discrimination of different classes. However, there are some statements, related to cork quality, that may be enumerated, namely: the number of defective spots; the presence of major defects; the presence of splits, notably those with vertical top to bottom runs, and the proximity of the defects to top edges. Having these facts in mind and using the two binary images *def(i,j)* and *maj(i,j)* , a number of features were measured. The measurement space may be divided into three main groups: a) bit quad features; b) projection features and c) vertical run length features.

A) *Bit Quad Features:* From a 2 x 2 binary pixel pattern (usually known as a bit quad [14]) one may have 16 different combinations. By computing the bit quad counts, a set of useful global features may be measured, such as the number of defective pixels and the Euler number. With such measurements, one has useful information about the presence and extension of important defects, the number of isolated defects in each image, the average extension of defects, etc. (Note that the Euler number is a good approximation to connected components count, particularly if they do not have "holes").

B) Projection Features: From the binary image *def(i,j)* we have defined two arrays with the projections into two orthogonal directions. With this information we are able, for instance, to characterize the proximity of defects to both edges, to evaluate its concentration and to detect the presence and extension of long vertical splits. Figure 16 shows the projections into two orthogonal directions of different quality samples.

C) Vertical Run Length Features: Bottle cork defects have a main development along the vertical direction, so that the characterization of vertical run lengths distribution may be useful for classification purposes. From *def(i,j)* we obtain a feature set with several measurements, weighting the influence of short and long runs, computing the number of vertical runs and characterizing the uniformity of the runs distribution.

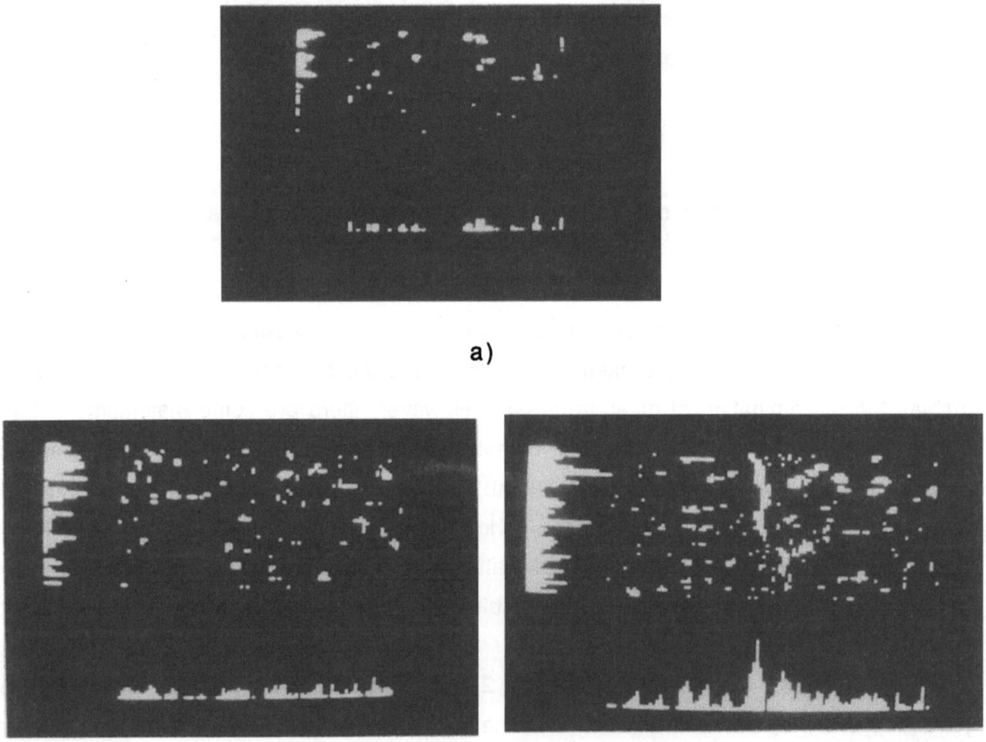

a)

b) c)

Figure 16. Vertical and horizontal projections.

The image preprocessing and analysis tasks are performed in SPAI according to the flow diagram of figure 17.

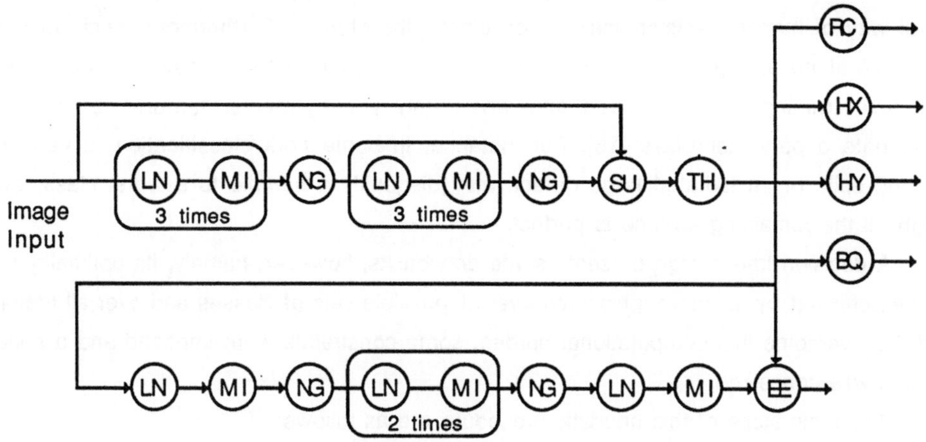

SPAI commands: LN - Largest Neighbour
MI - Maximum
NG - Negation
SU - Subtraction
TH - Thresholding
EE - Logical And
HX - Horizontal projection
HY - Vertical projection
BQ - Bit quad
RC - Run code

Figure 17. SPAI flow diagram for image preprocessing and analysis tasks.

6.2 Image Classification

Because of the complexity of this multivariate, multiclass problem which is mainly due to the absence of objective criteria (to discriminate the different classes) and to the low level of agreement among human inspectors, we have developed a statistical pattern recognition approach, to investigate the plausibility of the measurements extracted and the coherence of the feature space with the predefined "classes". If any evidence results from these first steps, the decision logic design will follow. We only describe the most successful approach, although SARF has been extensively used to evaluate many other alternatives.

In forming the training set, and in order to alleviate the well known low level of agreement among different inspectors, we have chosen a panel of three "experts" who have independently classified the same set of a few hundred bottle corks. We have confirmed about 40% of interobserver agreement and about 30% of total disagreement among all three. Thus, the six class training set used for decision logic design was formed by the 246 samples where a majority vote did exist.

In decision logic design we adopted an hierarchical approach, trying to replicate the natural way of human decision making, on classifying objects. Furthermore, decision trees have several advantages over conventional single stage classifiers, notably when fixed small size training sets are used and when only some special features are used to discriminate a particular class [16]. For instance, in bottle cork classification, a very thin and long split, or an isolated large defect, may disqualify a sample to a worst class, even though all the remaining surface is perfect.

A decision tree design presents some drawbacks, however; namely, its optimality may only be achieved by a thorough search over all possible sets of classes and over all features used. To overcome this computational burden, some constraints were imposed and a special heuristic was conceived.

The main steps of this heuristic procedure are as follows:

- generation of plausible dichotomies only, restricting to binary trees and to sets of neighbouring classes;
- for each two sets of classes settled in the previous step, feature selection was implemented using some useful criterion and an adequate search procedure (sequential procedures are usually used);
- selection of the dichotomy which "optimizes" the separation criterion used, and association to the dichotomy node the set of selected features and the decision function.

These three steps must be repeated until the decision tree is completed. Then we obtain a final hierarchical decision tree, with relevant information about its structure and tree nodes, namely: node nature (terminal or nonterminal node); linking graph among nodes; final classification associated with a specific terminal node; set of classes, feature set and decision function associated to nonterminal nodes.

Using this algorithm, a decision tree for bottle cork grading was designed. In figure 18. a) we can appreciate the resulting structure; the overall per node performance and the features used in each node are shown in figure 18. b). A feature set (three or four features) and a decision function (a Bayes classifier) are associated to each nonterminal node, |di|, of this decision structure . A final decision corresponds to each one of the six terminal nodes.

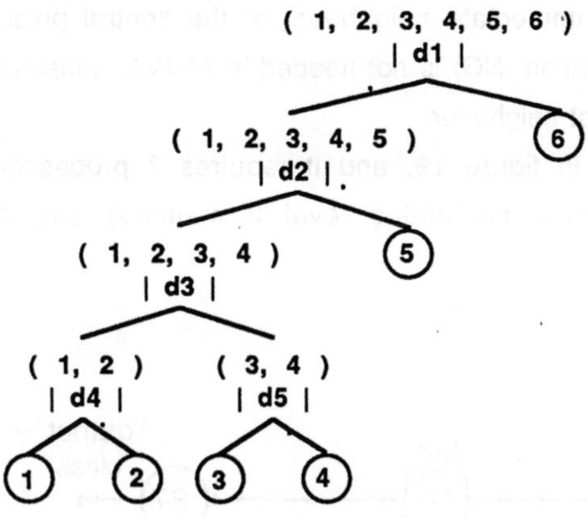

a)

Node	Features	C.Cl (di) (%)
d1	g5 g6 g7 r9	97.8
d2	g7 p4 r9	91.7
d3	g1 p11 p12	88.3
d4	g5 g6 r6 r9	79.2
d5	g1 g6 r5 r6	77.0

Legend: gi - bit quad global features
pi - projection features
ri - run length features
C.Cl(di) - % correct classification at node di

b)

Figure 18. Binary decision tree: a) structure; b) features and performance at a nonterminal node.

The overall performance of this decision tree was tested against the panel of "experts", and against each one of them. The global result is 70 % correct classification and about 60 % agreement with each inspector, which is much better than the 40% agreement between the members of the panel. Furthermore, we have only found 2.8% severe errors (class assigned is non-neighbour of correct class).

6.3 AMIVA System for Bottle Cork Inspection

The AMIVA system to implement bottle cork inspection must execute the pre-processing, analysis and classification procedures described in the previous sections.

The required performance specification is two corks per second; another important factor is a resolution of 50 x 100 pixels, which was found sufficient.

The primitive operations listed in figure 17 were separately executed in a single module in order to evaluate their execution speed at the required resolution; it was found that the execution times did not exceed 200 ms, except BQ which takes 330 ms; these results suggest that primitive operations can be grouped together, so that the module count is reduced; the limiting factor is a 500 ms operation time per module.

In grouping primitive operations, care has been taken to minimize the number of neighbourhood processors, which are required by the LN operation; this operation takes 161 ms, and, in fact, the same time holds for the sequence LN; MI, which is used in SPAI because LN evaluates the largest of the eight immediate neighbours of the central pixel, excluding it; on the other hand, the negation operation (NG) is not needed in AMIVA, because NP can compute the largest as well as the smallest neighbour.

The resulting AMIVA system is shown in figure 19, and it requires 7 processor submodules, 5 memory submodules (4 of them containing level 2 memory) and 4 neighbourhood processors.

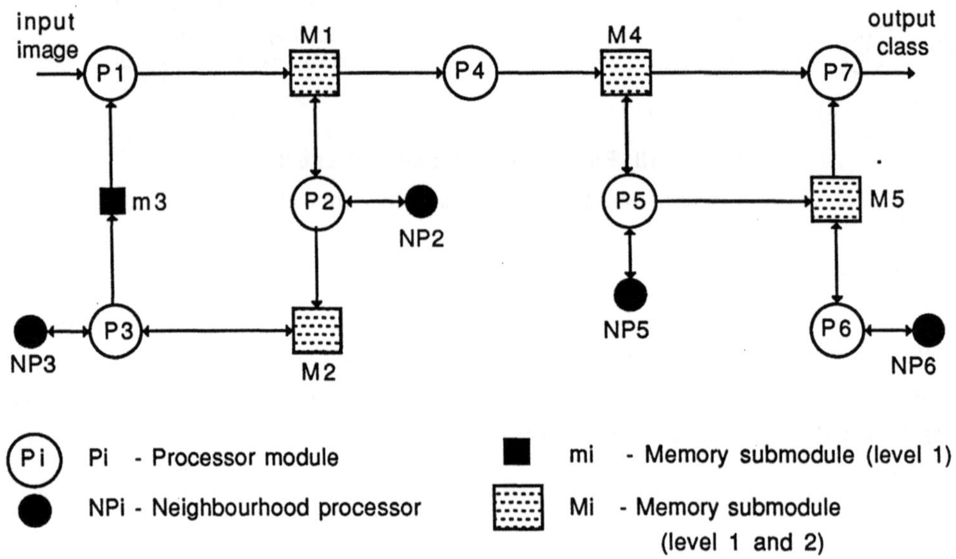

Figure 19. AMIVA system for bottle cork grading.

The first step was to group the operations of the streak spot detector in two modules (P2 and P3), each recirculating three times the image data, with a need for level 2 memory; the opening and closing operations were also grouped as two modules (P5 and P6).

The subtraction operation (SU) was assigned to a module (P1) that also inputs new images from frame buffers. It was also verified that this module could perform the thresholding operation (TH) and run-length coding (RC), as well as the computation of the associated features.

Another module (P4) was assigned the bit quad count, the computation of the corresponding features, and image and feature transport.

Finally, a module (P7) was assigned the anding operation (EE), the computation of major defect features (in fact, only area), and the decision logic to produce final classification.

The computation of the projections (HX and HY) and of its features was assigned to the module performing the opening function, which had a sufficient time margin.

The measured activity times for all processors submodules were:

P1	460 ms
P2 and P3	480 ms
P4	450 ms
P5	450 ms
P6	340 ms
P7	240 ms

Two phases were used in testing: in the first, processor P1, P2, P3 and P4 were assembled and tested; in the second phase, P4, P5, P6 and P7 were tested.

The latency time for phase 1 was about 760 ms; for phase 2, a latency time of about 490 ms was measured.

The expected results for a full system implementation are, then, a latency time of about 1.25 seconds and a maximum inspection rate of one cork every 480 ms.

7. Conclusions

A multiple microprocessor system (AMIVA) for automatic visual inspection applications was described. A close relationship with a software development package was maintained, so that the overall result is an integrated application development system that enables interactive algorithm testing and final prototype assembling.

The implementation of AMIVA systems requires the physical connection of a number of modules which are all identical, both in hardware terms and in software terms; this is an important feature, as it simplifies system maintenance and repair.

A special auxiliary processor was developed, in order to enable neighbourhood operations to be executed at a convenient speed.

A copy of AMIVA system software resides in every module, in a 2 Kbyte EPROM; this small system implements all the necessary inter-module communication procedures, so that the development of specific software for each module is an easy task.

The main goal in AMIVA design, namely to obtain systems with a real-time operation capability, with little effort and a short development time, has been fully achieved.

The automatic grading of bottle corks was discussed, as a demonstration example of the development cycle, from problem statement to prototype design.

Acknowledgements

The partial funding, by J.N.I.C.T. (Junta Nacional de Investigação Científica e Tecnológica) and by I.N.I.C. (Instituto Nacional de Investigação Científica), to the work herein described is gratefully acknowledged.

References

1. Fu, K. S.: Special Computer Architectures for Image Analysis; Robotica, vol. 2, Part 1, Jan. 1984.
2. Danielsson, P.E.: Designs for Parallelism in Image Processing; Univ. Linköping, Int. Rept. LiTH-ISY-I-0388, 1980.
3. Villers, P.: Present Industrial Use of Vision Sensors for Robot Guidance; in 12th Int. Symposium on Industrial Robots Proceedings, Paris 1982.
4. Holland, S. W.; Rossol, L.; Ward, M. R.: Consight-I: A Vison Controlled Robot System for Transferring Parts from Belt Conveyors; in Computer Vision and Sensor Based Robots; Dodd, G. G. and Rossol, L. (eds.); GM Research Laboratories, 1979.
5. Foith, J. P. et al.: Real Time Processing of Binary Images for Industrial Applications; in Digital Image Processing Systems; Bolc, Kulpa (eds.), Springer-Verlag, 1981.
6. Sternberg, S. R.: Parallel Processing in Machine Vision; Robotica, vol. 2, Part 1, Jan. 1984.
7. Persoons, E.: Industrial Applications of Machine Vision; NATO ASI on Industrial Robotic Vision, Leuven 1984.
8. Batchelor, B. G.; Cotter, S. M.: Inspecting Complex Parts and Assemblies; in Proc. 4th Int. Conf. on Robot Vision and Sensory Controls, London 1984.
9. Padilha, A. J.; Campilho, A. J.; Marques dos Santos, J. C.: An Interactive System for Image Processing and Analysis; in Proc. Portugal Workshop on Signal Processing and its Applications, 1981.

10. Batchelor, B. G.; Brumfitt, P. J.; Smith, B. D. V.: Command Language for Interactive Image Analysis; IEE Proc-E, Sept. 1980.

11. Campilho, A. J.; Marques dos Santos, J. C.: A Microcomputer Based System for Pattern Analysis and Recognition; in Proc. Portugal Workshop on Signal Processing and its Applications, 1981.

12. Monteiro, A. P.: Um Sistema Interactivo para Processamento de Comandos (Aplicado ao Processamento de Imagem); Master Dissertation, Faculty of Engineering of Porto University, 1987 (in portuguese).

13. Händler, W.: The Concept of Macropipelining with High Availabality; Elektronnishe Rechenenanlagen 17, 1973.

14. Pratt, W.: Digital Image Processing; John Wiley and Sons, 1978.

15. Batchelor, B. G.: Streak and Spot Detection in Digital Pictures; IEE Electronic Letters, vol. 15, pp. 352-353, 1979.

16. Argentiero, P.; Chin, R.; Beaudet, P.: An Automated Approach to the Design of Decision Trees Classifiers; IEEE Transactions on Pattern Analysis and Machine Intelligence, vol.4, pp. 51-57, 1982.

TACTILE SENSORS: APPLICATION ASSESSMENT FOR ROBOTIC HANDLING OF LIMP MATERIALS

Frank W. Paul
McQueen Quattlebaum Professor

Eric Torgerson
Ph.D. Graduate Student

Mechanical Engineering Department
Center for Advanced Manufacturing
Clemson University

ABSTRACT

Tactile sensors have a potential role in assisting the robotic manipulation of limp materials such as apparel fabrics, leathers, flexible composites, thin metals (foils), and polymer sheets. A review of the methods and principles utilized in existing tactile sensor technology is presented coupled with concepts for integrating tactile information with robot control. Possible uses of tactile sensors for limp material handling with robot manipulators are suggested. Research areas where further exploration and development is needed are identified.

Introduction

Harmon [1] defines tactile sensing as force sensing over an area using an array of sensor elements while touch sensing refers to binary (or gray-level) contact sensing at isolated points. Research efforts on tactile sensing have focused on developing methods for creating sensor arrays which can detect and measure forces between the sensor and some contacted object. No comprehensive study has been performed on the more difficult problem of integrating tactile information with the robotic execution of some specified task [2]. Experimentation

NATO ASI Series, Vol. F 64
Sensory Robotics for the Handling of Limp Materials
Edited by P. M. Taylor

on tactile sensor integration has been limited to simple touch sensors and heuristic control schemes for tasks such as robotic insertion of pegs into oversized holes.

Human tactile sensing, the basis for research in robotic tactile sensing, can be divided into two fundamental components [1]. The cutaneous component consists of the contact sensor arrays within and beneath the skin. The sensitivity of the skin sensors is particularly acute at the fingertips of the human hand. Cutaneous sensing is essentially static although human skin possesses some degree of compliance. The second component, the kinesthetic one, refers to the dynamic interaction of skin sensors with the environment via finger, hand, and arm movements. The hand/environment interaction reduces to the identification of objects within the environment and/or the application of forces to the environment. Humans identify objects without sight by actively "feeling out" their geometric properties. The hands are also used to apply forces in certain dynamic patterns such as those associated with the playing of a piano. The kinesthetic sense incorporates both these interaction types since they require the integration of tactile sensing with manipulator dynamics.

State-of-the-Art

Numerous types of tactile sensors ranging from simple touch to large arrays of intricate force sensors have been or are under development. Binary touch sensors such as microswitches are used extensively in assembly line applications for detecting the presence of a part at a particular location. The development of tactile arrays using different materials and force sensing principles has been restricted chiefly to the research lab. Methods for sensing and measuring forces produced by sensor/object contact include conductive materials, force transduction, optics, magnetics, and semiconductors.

Resistive and conductive materials whose electrical properties vary as some function of their deformation can be useful in tactile sensing. Force measurements can be derived directly or indirectly from the changes in electrical properties that occur during sensor/object contact. Most materials although low in cost suffer from hysteretic effects and low force sensitivity. Dario et al. [3] predict great potential for piezoelectric polymers due to their high sensitivity and ruggedness. Drawbacks of these polymers stem from their lack of DC response and their slippery surface which may cause object slippage.

An issue related to tactile sensing is the maintenance of a gripping force upon an object to prevent it from slipping within a robot end-effector. This issue is not truly kinesthetic since gripping force is normally static not dynamic. Gripping force can be measured by means of a force transducer. Force transduction implies the conversion of changes in electrical resistance, capacitance, and inductance into force measurements. Many robot end-effectors like the one developed by Parker [4] utilize strain gages to determine normal contact forces for acquisition of rigid objects. Spring-loaded potentiometers could similarly be used, but both these transducers are sensitive to hysteretic and environmental disturbances.

Fiberoptic tactile sensors have been developed by Jacobsen et al. [5] and the LORD corporation [6]. Figure 1 illustrates the principle behind fiberoptic tactile sensing. The sensor detects force by means of a light beam transmitted through a fiberoptic cable to a compliant material located at the sensor element. As the material deforms upon object contact, the intensity of the light transmitted back (reflected by a mirror)

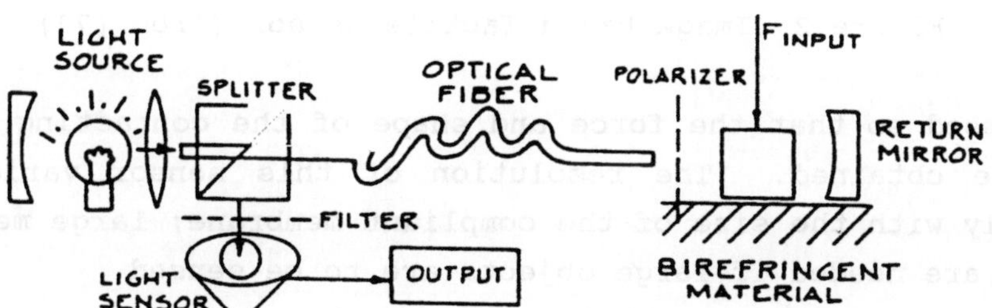

Figure 1 Fiberoptic Tactile Sensor (from [5])

through the cable changes because the deformed material blocks out certain light frequencies. Even though these fiberoptic sensors are free of hysteresis, the time response for acquiring tactile information is slow when compared with sensors using direct electrical connections.

Another sensor based on optical principles has been developed by Collins et al. [7] using a high-resolution (256x256) camera array chip. Figure 2 shows a representation of the image-based tactile sensor. The sensor consists of an image acquisition chip and lenses which direct an internal light source toward a compliant membrane situated at the contact surface. Visual image information from the deformed membrane is processed so that the force and shape of the contacting object can be obtained. The resolution of this sensor varies inversely with the size of the compliant membrane; large membrane areas are needed if large objects are to be sensed.

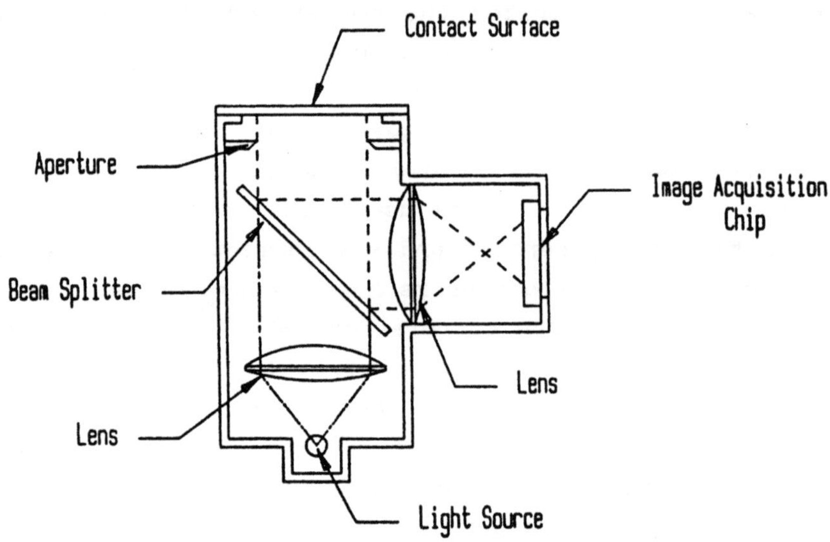

Figure 2 Image-Based Tactile Sensor (from [7])

Hackwood et al. [6] has developed a tactile sensor consisting of an array of magnetic dipoles embedded within a compliant medium. An illustration of a single sensor element is provided in Figure 3. Small electromagnetic sensors (solid line segments) situated beneath the dipole (segmented arrow) measure the position and orientation of the dipole as it is altered by distortions in the medium during object contact.

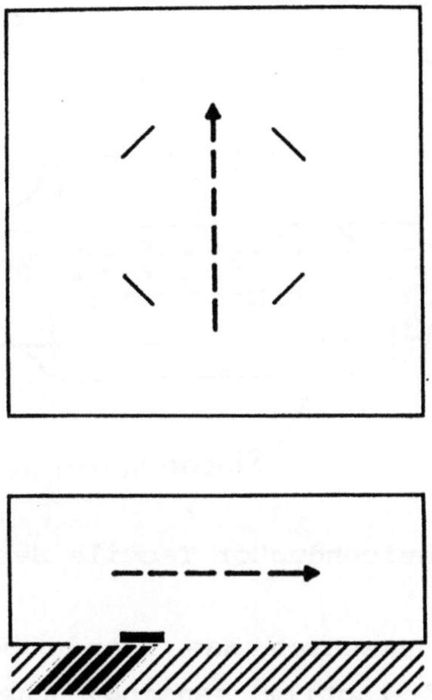

Figure 3 Top and Side View of Magnetic Tactile Sensor Element (from [6])

This discrete-magnet sensor is capable of determining normal and tangential forces as well as torques applied to the sensor surface. Hysteretic effects which are prevalent in magnets are reported to be eliminated by special placement of the sensor elements about each dipole.

An attractive solution to tactile sensing would be the incorporation of a tactile array within a small microchip which could implement parallel processing of all its sensor array elements. Raibert and Tanner [8] has achieved this on a few prototypes using VLSI techniques with elastic material bonded to the chip to protect it from the contact forces as shown in Figure 4. While no semiconductor tactile sensor has yet been introduced in the market, new developments in chip layout and operation software will make inexpensive, reliable semiconductor tactile sensors commercially viable within the next few years.

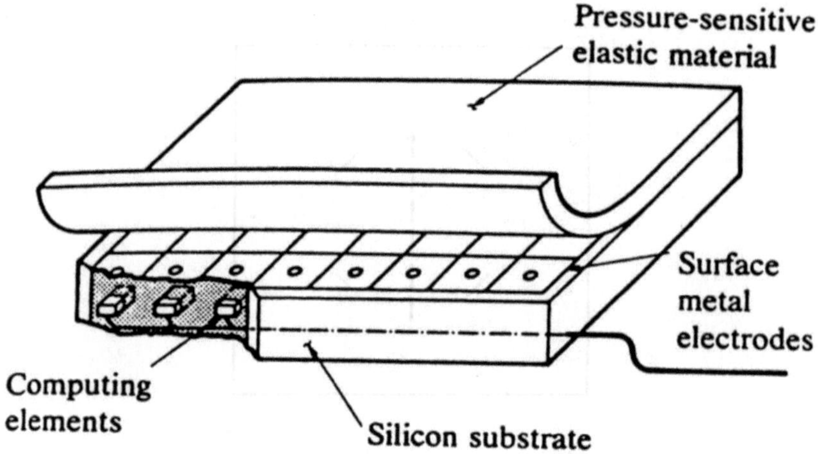

Figure 4 VLSI Semiconductor Tactile Sensor (from [8])

No significant research has been conducted on the kinesthetic element of tactile sensing; the human ability to perform intricate tasks and to identify complex objects without sight by the combination of tactile feedback from skin sensors with finger, hand, and arm motions is not understood at the conceptual level. The kinesthetic component must be studied since human recognition of and interaction with environmental objects is primarily achieved by dynamic manipulations involving tactile feedback. Investigations have been carried out on the similar issue of integrating force feedback from joint sensors to the control of robot arms and end-effectors. Raibert and Craig [9] studied the hybrid force and position control of simple manipulators. Jacobsen et al. [5] and Salisbury and Craig [10] investigated force control of articulated robot hands for grasping objects. The ad hoc control and integration schemes developed in these studies constitute a first step in the present understanding and future robotic application of the kinesthetic sense.

Application Areas in Limp Material Handling

Current applications of tactile sensing in limp material handling focus principally on the cutaneous aspect or the static sensing of contact forces. Three cases where tactile sensing plays an active role have been identified in the literature. Gershon and Porat [11] integrated a cloth tension (tangential force) sensor with a robotic sewing system to maintain tension on fabric as it undergoes sewing. Maintaining tension during a stitching operation is necessary to prevent the cloth from buckling and binding underneath the sewing head. Despite the novel use of a tension sensor, the robotic sewing system was reported to experience buckling problems when stitching along sharp curves at the fabric edge. Kemp et al. [12] installed a simple touch sensor on their two-fingered fabric gripper for determining the position (height) of a fabric stack prior to the gripper's acquisition of the top piece off the stack. Strain and semiconductor pressure gages were used by Okabe et al. [13] to sense cloth material thicknesses by means of a two-fingered robot gripper.

Tactile sensing appears to have a limited but extremely useful potential in limp material handling. Limp materials are normally flat and thin with areas of varied size and shape. Their three-dimensional shape is generally dependent on the method of support and is unpredictable and unstable. Identification of individual fabric pieces is ordinarily accomplished by recognizing differences in shape and size as well as pattern and/or color. These properties are distinguishable through visual not tactile sense. Similarly, the position and orientation of limp material pieces are visual (spatial) features. Object identification via tactile sensing is better associated with recognition of rigid parts with stable three-dimensional geometries. However, most tasks involving limp material manipulation require tactile sensing especially those involving complex three-dimensional adjustments in the material's geometry. For instance, a human operator could perform a turning operation on a collar or cuff without visual assistance solely by the use of his/her hands. The operator would need an acute kinesthetic sense to carefully reshape the collar or cuff by precise finger and hand motions; tactile feedback would continuously sense the changing geometry of the collar or cuff during the turning process. To perform this type of operation with a robot end-effector with tactile sense would be difficult but technically feasible.

A practical application of tactile sensing as suggested earlier is the determination of the amount of tension (or tangential force) on limp material undergoing stretching. Tension sensors would be useful in fabric stitching to prevent fabric puckering under the needle and in stretching out limp material pieces for accurate robotic placement on to two-dimensional molded surfaces. The lay-up of flexible composite sections and the edge alignment of matching cloth pieces are two such instances of robotic limp material placement. Maintaining tension in limp materials would prevent their wrinkling and buckling during positioning. Slip detection of limp materials such as apparel fabric due to gravitational force is generally not required because of most material's low weight. Figure 5 shows a view of two robotic grippers holding a limp material section in tension. The tensile force distribution acting upon the gripper is illustrated in the figure along with those forces acting parallel to the section plane. The position adjustment and grasping force of the grippers would vary with the material's thickness and the tensile force distribution placed on each gripper. Gripper clamping force must be increased to prevent lateral slipping of the material as it is stretched by the diametric pull of the end-effectors.

The measurement of normal contact forces is not as important a consideration for tactile sensing of limp materials as it is for rigid objects. Unsupported limp materials when contacted by a force sensor would impart no significant normal forces to the sensor due to the low (nearly zero) stiffness of the material. The detection and measurement of normal forces may have application, however, in operations involving heavy-weight materials with relatively high weights and stiffnesses such as tent and upholstery fabrics. The benefit of tactile

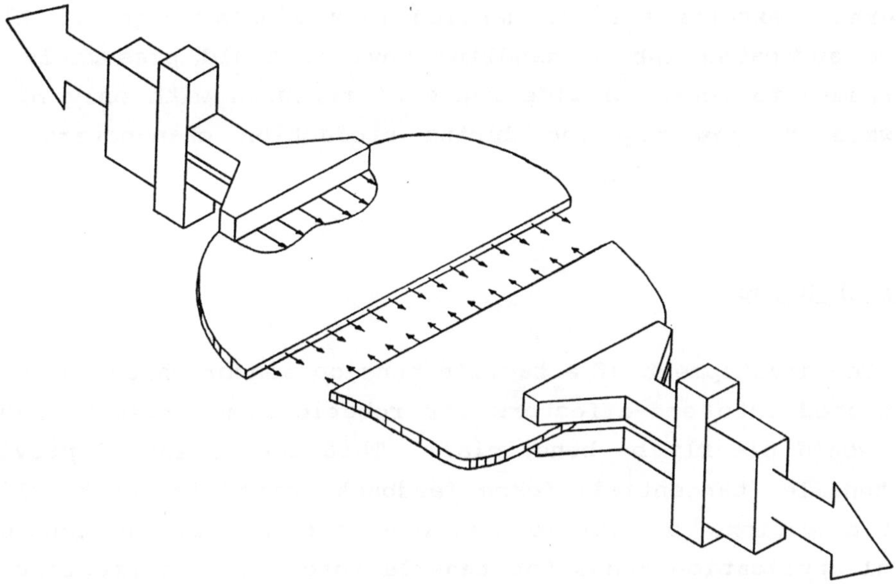

Figure 5 Two Robotic Grippers Holding Limp Material

sensors for flexible materials derives chiefly from the determination of the absence or presence of the material within the end-effector. Knowledge of the geometry of an acquired fabric section or flexible composite relative to the robot end-effector is more useful than the evaluation of the normal contact forces imposed by the material upon the end-effector. An example of this fact is the robotic placement of a fabric section to some specified location. The three-dimensional spatial geometry of the section relative to the robot end-effector must be known before the fabric section can be manipulated to conform to the final, desired geometry. The value of the normal force distribution imposed by the section on the end-effector's tactile array may not provide useful information for this task.

Material texture is one property that could be quantitatively defined by some sort of tactile sensor. The sensor could measure the tangential force imparted to it as fabrics of various textures were moved over the sensing element. The need for this sensing capability is not clear since fabrics with different textures are all handled in a similar manner. Also future automated fabric handling devices could presumably be programmed to accept a wide range of textures with only minor software changes required during production changeovers involving material textures.

Research Needs

The development of a tactile tension sensor which could be integrated into end-effectors for robotic limp material handling would clearly be beneficial. This sensor should provide the tensile (tangential) force feedback needed for controlling robotic motions as fabric sections are placed in tension. Useful application areas for tensile force sensing are robotic sewing and edge alignment during fabric assembly. A tactile array that could detect the presence/absence of a limp material ply of various thicknesses should also be designed and implemented. The sensor elements in this array may not need to output a continuous signal for measuring force but only a binary one indicating where in the array the fabric piece is touching. This type of sensor could assist in determining the spatial geometry of limp materials held by robotic devices. Both of the suggested tactile sensors must be compact and relible enough to be utilized in an industrial setting.

Fundamental research is required regarding the kinesthetic issue of tactile and vision sensor integration with control schemes for the robotic manipulation of limp materials. Issues such as sensor fusion, kinematics, force control, dynamic sensing, and task completion must be addressed. Studies should also be undertaken on the design of robot end-effectors for limp material handling appplications. The use and incorporation of tactile and visual sensors and the basic end-effector structure (i.e. anthropomorphic, two-fingered, etc.) constitute important criteria in these studies. The design of limp material handling end-effectors may prove to be entirely dependent upon the tasks and materials they are designed to address.

References

1. Harmon, L. D., "Tactile Sensing for Robots," Chap. 10, Recent Advances in ROBOTICS, Ed. by G. Beni and S. Hackwood, Wiley-Interscience, New York, pp. 389-424.
2. Henderson, T.C., P. Allen, I. Cox, A. Mitiche, H. Durrant-Whyte, and W. Snyder, "Workshop on Multisensor Integration in Manufacturing Automation," NSF Sponsored Workshop, NSF Grant No. DMC-8612180, Feb. 1987.
3. Dario, P., C. Domenici, R. Bardelli, D. De Rossi, and P.C. Pinotti, "Piezoelectric Polymers: New Sensor Materials for Robotic Applications," 13th Int. Symp. on Ind. Robots, pp. 14-34-14-49.
4. Parker, J.K, "Characterization and Control of Object Acquisition using a Robot Hand," Ph.D. Dissertation, Mechanical Engineering Dept., Clemson University, May 1985.
5. Jacobsen, S.C., J.E. Wood, D.F. Knutti, and K.B. Biggers, "The UTAH/M.I.T. Dextrous Hand: Work in Progress," The Int. Journal of Robotics Res., Vol. 3, No. 4, 1984, pp. 21-49.
6. Hackwood, S., G. Beni, L.A. Hornak, R. Wolfe, and T.J. Nelson, "A Torque-Sensitive Tactile Array for Robotics," The Int. Journal of Robotics Res., Vol. 2, No. 3, 1983, pp. 46-50.
7. Collins, A.S., G.G. Hastings, and F.W. Paul, "An Optical Tactile Sensor for Manufacturing Applications," Sensors for Manufacturing, Presented at the Winter Annual Meeting of the ASME, Dec. 1987, Boston, Mass., pp. 109-119.
8. Raibert, M.H., J.E. Tanner, "Design and Implementation of a VLSI Tactile Sensing Computer," The Int. Journal of Robotics Res., Vol. 1, No. 3, 1982, pp. 3-18.
9. Raibert, M.H. and J.J. Craig, "Hybrid Position/Force Control of Manipulators," Trans. of the ASME Journal of Dynamic Systems, Measurement, and Control, Vol. 102, June 1981, pp. 126-133.
10. Salisbury, J.K and J.J. Craig, "Articulated Hands: Force Control and Kinematic Issues," The Int. Journal of Robotics Res., Vol. 1, No. 1, 1982. pp. 4-17.
11. Gershon, D. and I. Porat, "Vision Servo Control of a Robotic Sewing System," IEEE Int. Conf. on Robotics and Automation, 1988, pp. 1830-1835.
12. Kemp, D.R., G.E. Taylor, and P.M. Taylor, "An Adaptive, Sensory Gripper for Fabric Handling," Proc. of the 4th Int. Symp. on Robotics and Automation, Amsterdam, 1984.
13. Okabe, H., E. Ono, H. Akami, and K. Sasaki, "On the Sensing of Cloths by Sensors Implanted in Robot Fingers," Bulletin of Research Institute for Polymers and Textiles, Japan, 1985, No. 147, pp. 23-29.

Round Table Discussion
Sensing: Trends and Future Research

Chairman: Dr. P.E. Wellstead
Secretary:Mr. D.G. Caldwell

The chairman opened the discussion by outlining the sensing systems mentioned during today's session (vision and tactile) and leaving room for discussion of some as yet unconsidered sensors. The debate was started by asking "Is vision ready for use in industry and what developments are required?"

It was proposed that a colour vision would be very useful in many matching operations. This was generally felt to be an over complication and that grey levels would be sufficient in most instances, and it was actually disclosed that preliminary tests and prototype models had been used to match checks and stripes. Despite this a few industrialists felt that colour will still be an important development.

It was believed that, at present, industry was willing to use vision systems but as these needed to be reliable and robust only simple designs were considered to be viable at present. It was stressed that the clothing industry was replacing a low-cost, undervalued system, the operator, and it was difficult to produce a vision system which was economically acceptable even allowing for special insurance costs occurring in vision operations (ie compensation for damage to eyesight).

The chairman now directed the discussion towards the issue of 2D versus 3D vision, asking "Is 3D vision necessary given that most textile problems occur in two dimensions?".

The debate seemed to suggest that the major motivation for using 3D vision was that the systems were available and were often used when not really necessary. At the same time it was conceded that there were certain specialist activities for which 3D vision was at present the most feasible solution.

Dr. R. Kelley suggested that in general vision and sensor technology were solutions looking for problems, and asked "What are we trying to achieve using this technology?". This led to a discussion of the uses of more complex systems and suggested that these were primarily targeted at research and problem analysis rather than being the cost optimal solution. Allied with this need to find the right applications for sensor technology was a feeling that the range of devices available off the shelf was very limited, leading to imperfect and overly complex solutions.

Consideration of other non-vision optical systems revealed that these systems were perceived to complement standard vision technology rather than competing with it and as such there were areas in which it would be superior to present visual techniques.

Concluding the debate on vision the chairman suggested that colour research may be useful although not absolutely necessary, while control and fault checking were areas in which vision could find useful applications.

Moving on to tactile sensors it was asked "What are the tactile requirements?".

NATO ASI Series, Vol. F 64
Sensory Robotics for the Handling of Limp Materials
Edited by P. M. Taylor

The salient problems with tactile technology were seen to be poor resolution, hysteresis and lack of robustness. For the garment industry it was accepted that perhaps development of tangential force sensors would be an advantage since material can be more easily handled vertically than horizontally. The continuing discussion was rather disjoint and it was suggested that this was due to a lack of definition of 'tactile'. It was felt that rather than one sensor, 5 or 6 different senses were being considered which obviously made focusing research difficult. It was concluded that a clarification of the objectives and sensing requirements of tactile systems was of great importance.

Finally the chairman asked "Are there any aspects of sensor technology which needed consideration or had not yet been addressed?"

It was proposed that three factors needed more consideration. Firstly temperature sensors, although this may be attacked as a tactile sensing problem. Secondly it should be noted that most sensors give static outputs while most natural sensing systems only produce a dynamic output, reducing processing by only considering signals when a change occurs. Finally it was felt that the siting of the sensors and the problems of fusing the information required more attention before reliable robotic sensing would be possible.

CHAPTER IV

PLANNING AND CONTROL

TRAJECTORY PLANNING AND OBSTACLE AVOIDANCE

Bernadette Gasmi - Claude Reboulet
CERT - DERA, Toulouse, France

ABSTRACT:

In this paper we enumerate some requirements necessary for a task level language. These are:

* How to generate safe paths among obstacles for moving the robot and the object carried, from their initial configuration to the final one.

* How to compute grasping positions of given objects.

* How to generate fine motions that require sensor interaction.

An algorithm is presented which deals with translating an elliptic object among elliptic obstacles, without collision.

This paper reviews several geometric trajectories required in a robot programming system.

KEYWORDS: Robot programming language; trajectory planning; collision avoidance; compliant motion.

NATO ASI Series, Vol. F 64
Sensory Robotics for the Handling of Limp Materials
Edited by P. M. Taylor

I) INTRODUCTION

Different approaches for developing an ideal robot programming language have already been proposed.
The robot languages are generally divided into four classes according to the level of detail in which the tasks the robot has to perform are expressed : joint level, manipulator level, task level and objective level.

At joint level, the description of a task is expressed in terms of the control commands required to move each motor.
With a manipulator level, the tasks to be performed are described in terms of motions of the robot end-effector. The programmer has to specify the successive locations for the frame connected to the robot end-effector.The programmer must be able to reason quantitatively in the three dimensional world and has to specify a level of geometric detail; this is very difficult for a human to do.
With the task level, the task is described in terms of operations on the objects.Task level programming systems require complete geometric models of the environment and of the robot. However, the programmer must choose the sequences needed to complete the different subtasks and also decide on the use of sensors .
In objective level programs, the user specifies only the desired final result.

A task level language provides several interesting features :

1) The off-line programming aspect : robot programs are developed without requiring the use of the robot itself.It allows the robot to remain in the workcell while the next task is being programmed. Off-line programming systems, however, need a geometric modelling system.
2) Flexibility is an important feature as it allows changes to or the addition of elements in the cell, thus modifying its spatial organization. As a robot program is stored at task level, positions are saved relative to the positions of objects in the workplace. The locations of these reference objects and of the robot itself in the workcell may thus be changed without invalidating the programs.
3) The programmer need not worry about the details of numerical operations. He does not need to express the successive transformations that have to be applied to relate a frame to the absolute reference frame. The programmer reasons in terms of the objects being manipulated. The numerical information is automatically updated after every manipulative action .

In order to offer the programmer the possibility of concentrating on the logical aspects of programming a task, a task level language must include certain basic elements. The elements taken into consideration are:

*How to grasp an object

*How to avoid obstacles

*Fine motion strategy

All of them require a structured geometric database.

This gives the robot programming system:

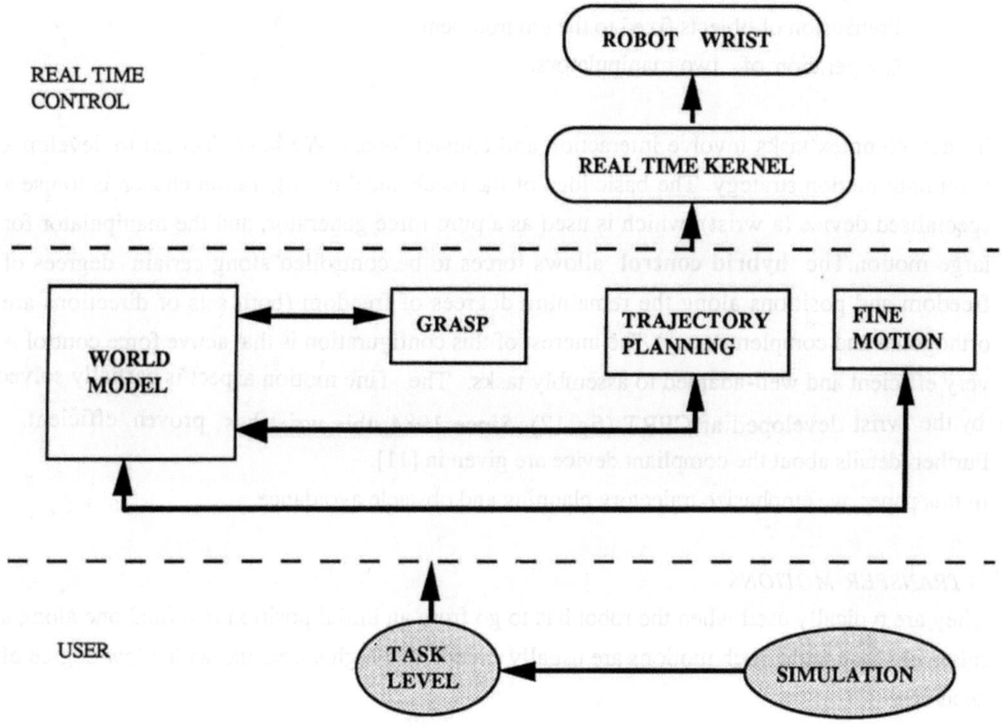

ROBOT PROGRAMMING SYSTEM

We have decided to start with the design of a manipulator-level system.

II) TRAJECTORY PLANNING

A) MOTIONS

Before examining this level, it is important to mention that, in robotics, there are basically two different types of motion :

1) *FINE MOTIONS*

They are needed for assembling pieces, part mating or grasping. They involve short trajectories in a highly constrained environment. The speed is limited, the degree of accuracy high. Combined strategies of force and position control should be used to achieve the desired final configuration.

The task level language we are developing must take into account all the capabilities of the wrist developed in CERT. It is specialised in tasks with contact.

We find tasks with contact in :

- Assembly tasks
- Manipulation of tools on a surface
- Prehension of objects fixed to the environment
- Cooperation of two manipulators.

In fact, complex tasks involve interaction and contact forces. We have decided to develop a compliant motion strategy. The basic idea of the mechanical configuration chosen is to use a specialised device (a wrist), which is used as a pure force generator, and the manipulator for large motion.The **hybrid control** allows forces to be controlled along certain degrees of freedom and positions along the remaining degrees of freedom (both sets of directions are orthogonal and complementary).The interest of this configuration is that active force control is very efficient and well-adapted to assembly tasks. The fine motion aspect is partially solved by the wrist developed at CERT (fig 12). Since 1984, this wrist has proven efficient. Further details about the compliant device are given in [11].

In this paper, we emphasize trajectory planning and obstacle avoidance.

2) *TRANSFER MOTIONS*

They are typically used when the robot has to go from an initial position to a final one along a relatively long path. Such motions are usually executed at high speed and with a low degree of accuracy.

The set of motion instructions constitutes the most obvious aspect of robot programming. There are many different methods which are employed to generate the path that the manipulator follows in going from the current configuration to the final one. We can classify them according to the type of path along which the tool tip moves.

a) JOINT MOTION

Joint motions are generated by interpolating each joint's control variable from its initial value to its desired value. Each path point is specified in terms of a desired position and orientation of the tool frame {T} relative to the station absolute frame {S}. Each of these points is "converted" into a set of desired joint angles by application of the inverse kinematics.Generally, all the joint variables are interpolated in such a way that the joints complete their motions simultaneously. Joint motion is used in the large motion parts of manipulator trajectories when the manipulator is moving clear of any object and when minimum time is of importance.

b) CARTESIAN MOTION

The spatial shape of the path taken by the end-effector following a straight line in joint space is a complicated shape in Cartesian space, which depends on the particular kinematics of the manipulator being used. In this section we describe several Cartesian coordinate motions. The advantage of Cartesian motion as compared to joint motion is that motion between trajectory segment end points is well-defined. In the following section we introduce some Cartesian motions .

The motion is generally achieved by means of a straight line translation and a rotation about some fixed axis in space. This occurs when the manipulator is performing a precision insertion and the required relative motion between the manipulator and the external workpiece is a straight line. If we can find such a line and axis, then we can produce a motion of controlled linear and angular velocity which would, for example, enable us to move the tool frame from its current value $\{R_i\}$ to a desired final value $\{R_f\}$.

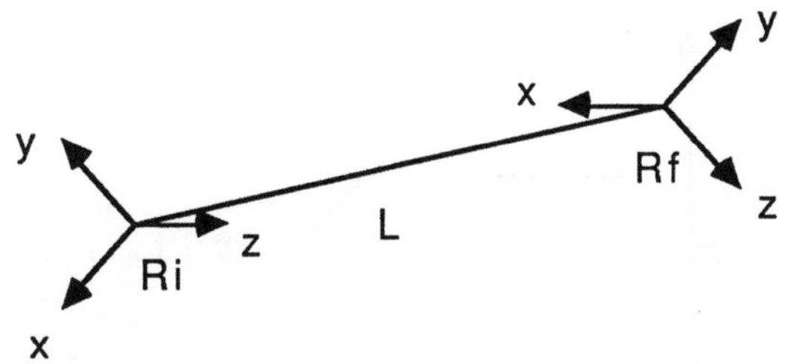

One of the simplest ways to generate this motion corresponds to the application of positive acceleration a for half the motion and - a for the second half of the motion (fig 1a). If the maximum velocity is reached, we have a part at constant velocity (fig 1b).
The movement on this type of trajectory is what is called a "Bang-Bang Law".

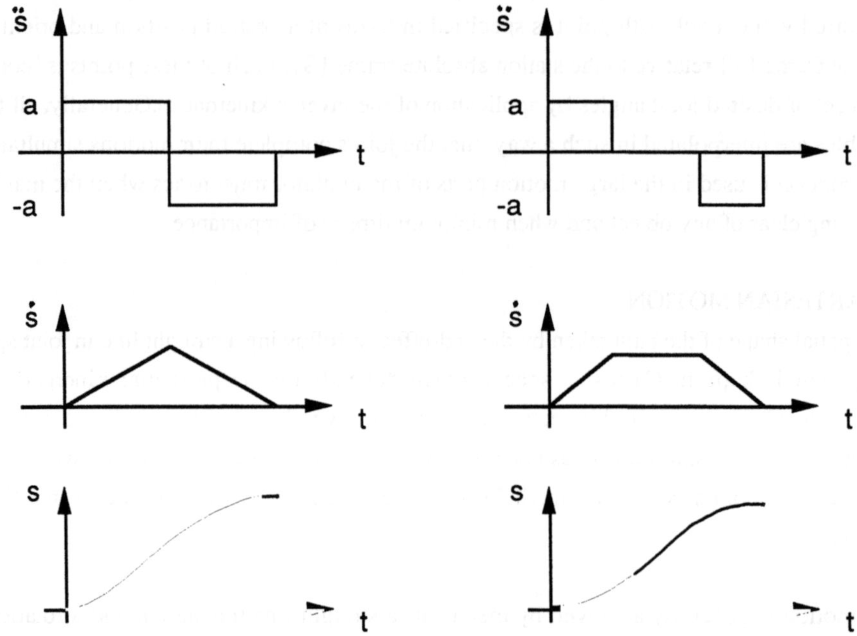

Motion trajectory

fig 1a fig 1b

If we want a coordinate motion between translation and rotation, we shall have to satisfy the most constraining law.

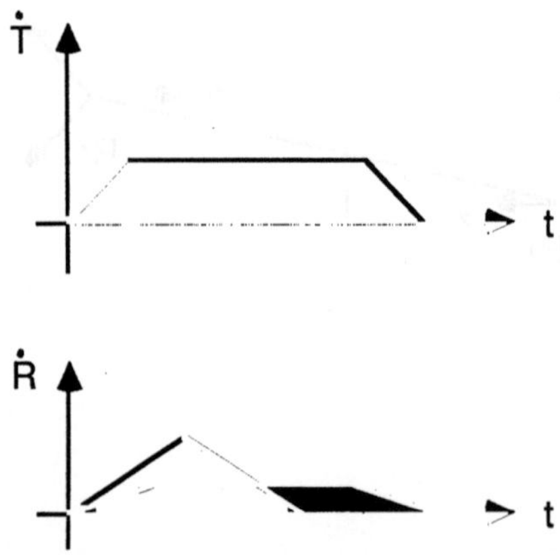

In most applications, it is not satisfactory to specify a robot motion by means of its initial and final position. Generally, paths are commonly specified by indicating a sequence of intermediate positions, known as via-points, that the robot should traverse between the initial and final positions. It must be decided whether the motion has to pass exactly through the specified via-points and what magnitude of deviation may be tolerated. Each of the via-points is actually a frame which specifies both the position and the orientation of the tool relative to the absolute reference frame. Straight lines connect via-points and parabolic blend regions are added around each via-point.

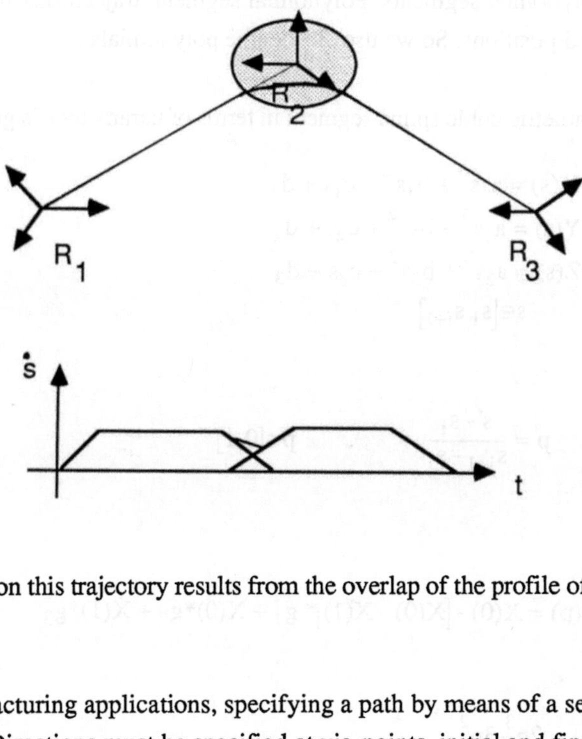

The movement on this trajectory results from the overlap of the profile of the velocity of the two straight lines.

In some manufacturing applications, specifying a path by means of a sequence of via-points is not sufficient. Directions must be specified at via-points, initial and final positions. Usually, it is desirable for the motion of the manipulator to be smooth. We define a smooth function as one which is continuous and has a continuous first derivative. A single example of these requirements arises when grasping parts : the robot cannot approach the grasp point from arbitrary directions (for example, an object fixed in a vice).

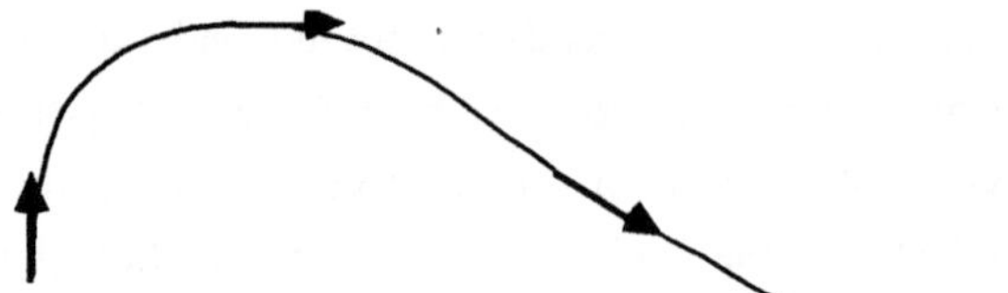

Following a prescribed path in an ideal manner means a one DOF motion represented by path coordinate s with $0 \le s \le sf$. The trajectory of each Cartesian point is specified by several smoothly connected polynomial segments. Polynomial segment trajectories are determined to pass through the selected positions. So we use 3rd degree polynomials .

The equations for a parametric cubic spline segment in terms of parameter s is given by :

$$X(s) = a_1 s^3 + b_1 s^2 + c_1 s + d_1$$
$$Y(s) = a_2 s^3 + b_2 s^2 + c_2 s + d_2$$
$$Z(s) = a_3 s^3 + b_3 s^2 + c_3 s + d_3$$
$$s \in [s_i, s_{i+1}]$$

with the change :

$$p = \frac{s^{\bullet} - s_i}{s_{i+1} - s_i} \qquad p \in [0,1]$$

we finally get :

$$X(p) = X(0) - [X(0) - X(1)] * g_1 + \dot{X}(0) * g_2 + \dot{X}(1) * g_3$$

with

$$g_1(p) = 2p^3 - 3p^2$$
$$g_2(p) = p^3 - 2p^2 + p$$
$$g_3(p) = p^3 - p^2$$

and $\dot{X}(0)$ direction at s_i

$\dot{X}(1)$ direction at s_{i+1}

Y(s) and Z(s) are calculated in the same way.
Then we use a profile of velocity on p.

The last primitive, "Move-with-force", is very important. It requires trajectory planning and fine motion. Motion during an assembly operation may have up to three submotions : a compliant departure from the current position, a free motion towards the intermediate position of the task step and a compliant motion to reach the goal position.

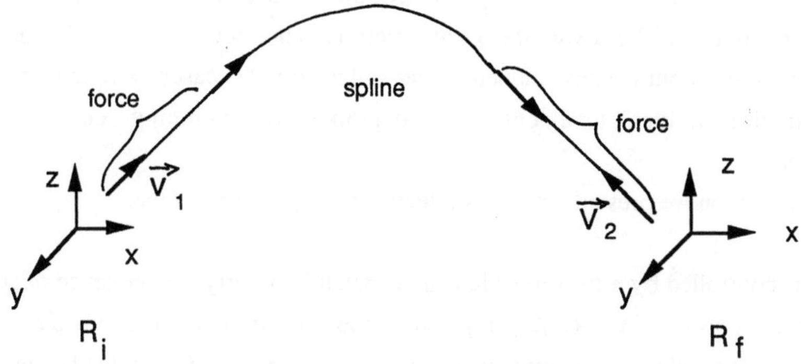

First we apply a force about axis V_1, using hybrid control (V_1 force controlled and the complementary directions position controlled). Then we switch hybrid control to position control without stopping, join these two points with cubic splines and we return to hybrid control to carry out the final task.

B) OBSTACLE AVOIDANCE

The central problem when programming at task level is collision avoidance. This means planning trajectories that are as short as possible and are safe for the object being carried and the manipulator. Path planning and obstacle avoidance algorithms employ a multitude of different representations and use different strategies.

Among global strategies we find :

◊The VORONOI diagram used in many two-dimensional approaches [8].

◊The objects and obstacles are assumed to be polygons. The approach is based on shrinking object A to a point and increasing the size of the obstacles to compensate for the size of the original object. A map of safe straight line paths between expanded obstacle vertices, called a visibility graph, is then created. The minimum distance path can be found in the visibility graph [5].

◊The octree model [9] divides the space into square regions.Each region is either empty or contains an obstacle. In the event of an obstacle being in a region, the region is further divided into smaller regions for greater accuracy.

Among local strategies we find :

The method of potential fields. Khatib [10] describes a potential field around an obstacle that repels the robot and one around the goal that attracts the robot.
The difference between global and local strategy is that in local strategy we use local information, with few details but rapidly acquired, and a strategy is defined to bypass the obstacle encountered.The advantage of this method is that it can run in real time, but as it has local information about the environment, it can fail in complex cases like dead-ends.
A combination of local level control with global level planning seems to be the most appropriate.
In the next section we present a mixed strategy which runs in real time.

The cell is controlled by a system which can constantly modify the sequence of activated tasks or modify the grasp or the assembly of pieces in the event of a failure of one element of the cell. It must be able to adapt to dynamic changes, so trajectory planning must be set in a real time context.

The problem can be formulated as follows "Given an object A and a set of obstacles $\{E_j\}$, all within the workspace R, what is the trajectory that is to be followed by A so that it is moved from its initial configuration to its final one in such a way that it remains within the workspace and does not collide with any of the obstacles?".

1) *HYPOTHESIS*

In this paper, we shall consider the case when both the object and the obstacles are modeled by a cylinder with an elliptic base and when the only motion permitted is translation. The position of all objects and obstacles is known but can change between two motion commands.
We must keep in mind that this strategy is well adapted to a SCARA type robot.

To simplify the notation, we shall refer to an ellipse E as a quintuple $E = (h,k,a,b,\alpha)$ where

1)h and k are the x and y coordinates respectively of the center of the ellipse,
2)a and b are the lengths of the semi-major and semi-minor axes respectively,
3)α is the angle of the orientation of the major axis of the ellipse.

We also use the standard equation :

$$ax^2+by^2+cxy+dx+ey+f=0$$

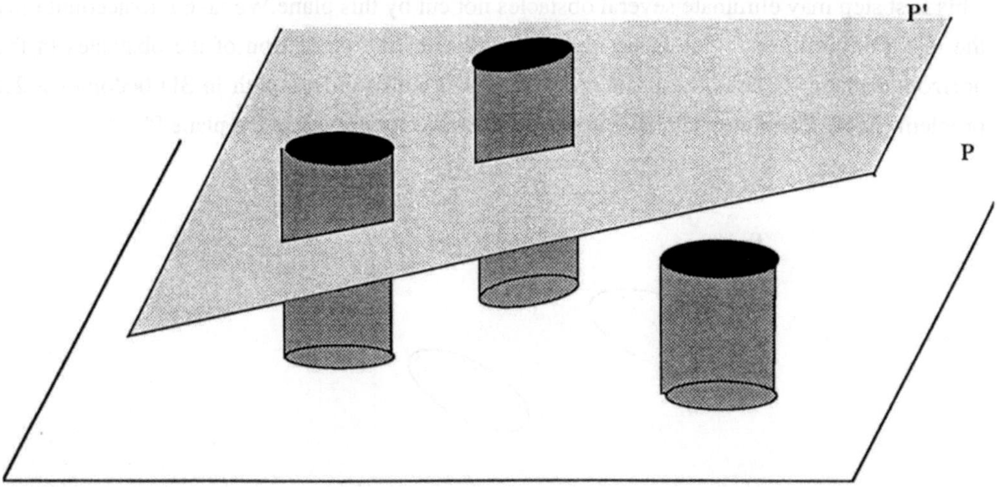

fig 2

2) *SOLVING THE PROBLEM*

With this particular configuration (fig 2), we first oblige the trajectory to remain in a given plane. This plane contains the initial and final positions. Among the infinity of planes passing through A,B we consider P' to be the one which makes the smallest angle with the horizontal plane P (fig 3).

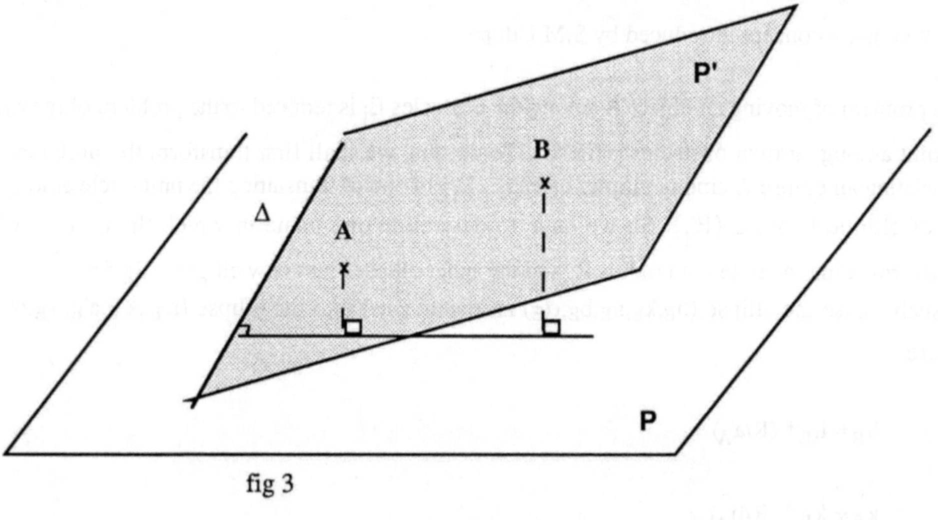

fig 3

This first step may eliminate several obstacles not cut by this plane.We take into account only the remaining ellipses . The basic idea is to consider the projection of the obstacles in the horizontal plane P. Thus the problem of finding a collison-free path in 3D becomes a 2D problem (fig 4). Then we shall have to project the trajectory found in the plane P'.

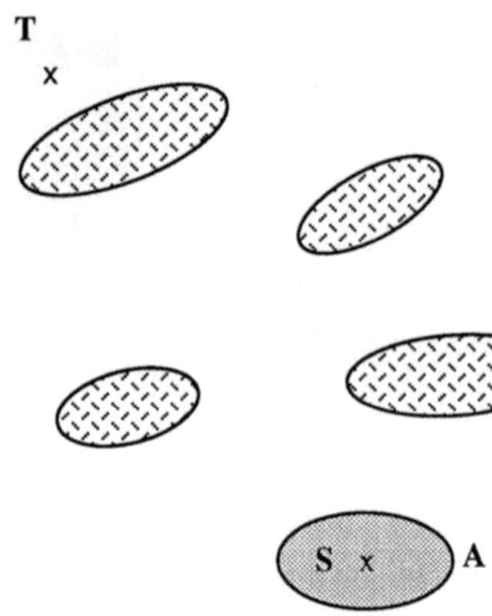

fig 4

We then use a concept introduced by S.M Udupa :

The problem of moving an object A among the obstacles E_j is reduced to the problem of moving a point among "grown obstacles" (fig 5a). To do this, we shall first transform the problem of translating an ellipse A among elliptic barriers $\{E_j\}$ to one of translating the unit circle among a set of elliptic barriers $\{E'_j\}$. So we have to determine one plane in which the ellipse A is transformed into a circle with radius R, and the other ellipses into new ellipses (fig 5b).
In such a case the ellipse $(h_E,k_E,a_E,b_E,\alpha_E)$ is transformed into the ellipse $(h'_E,k'_E,a'_E,b'_E,\alpha'_E)$ where,

$$h'_E = h_E * (R/a_A)$$

$$k'_E = k_E * (R/b_A)$$

$$a'_E = a_E\, R((\cos\alpha_E/a_A)^2 + (\sin\alpha_E/b_A)^2)^{\frac{1}{2}}$$

$$b'_E = b_E\, R((\cos\alpha_E/b_A)^2 + (\sin\alpha_E/a_A)^2)^{\frac{1}{2}}$$

$$\alpha'_E = \mathrm{Tan}^{-1}(\mathrm{Tan}\alpha_E(a_A/b_A))$$

The ellipse $(h_A,k_A,a_A,b_A,0)$ is transformed into the circle $(Rh_A/a_A,Rk_A/b_A,R,R,0)$.

Remark :
These transformations, being linear, map straight lines into straight lines.

Then we increase the size of the ellipses with the radius of the circle. Even though these new figures are not strictly ellipses,we can consider them as such, with a new major axis (a+R) and a new minor axis (b+R).

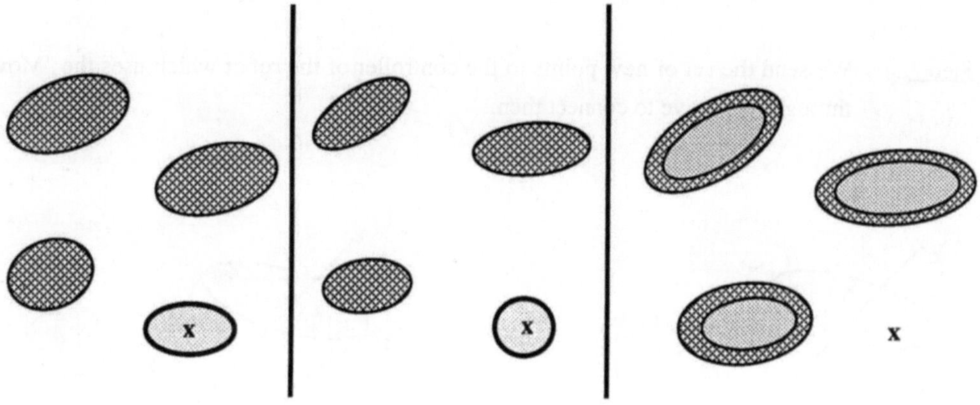

fig 5a fig 5b fig 5c

So, the problem is to move a point among obstacles (fig5c).Source S and target T are specified.

3) *SOLVING THE FINDPATH PROBLEM*

<u>Step 0</u> : We classify the ellipse nearest to the target in an array Tab.

Step 1 : Join the points S and T by a straight line and check if there are ellipses on this path (fig 6a). The first found in Tab stops this step. Go to step 2.

Step 2 : Tangent the ellipse. As we find 2 tangents we choose that with the smaller angle (fig 6b). We get an intermediate point T_1. Go to step 3.

Step 3 : Check if the point is in another ellipse. No:go to step 4.Yes: tangent this new ellipse and take the tangent which is not contained by the first ellipse tangented (fig 7). Go to step 4.

Step 4 : Go to step 1 until there are no more ellipses on the straight line. Go to step 5.

Step 5 : The last intermediate target is taken for the new source .Memorise it. Go to step 1 until there are no more ellipses on the straight line. Go to step 6.

Step 6 : At this step, we get a set of points (fig 8). We must now join them, so we have to compute the common tangent to two ellipses (fig 9). We use the Newton-Raphson method for identifying these points.We can find it in the appendix.

Step 7 : We send the set of new points to the controller of the robot which uses the "Move through" primitive to connect them.

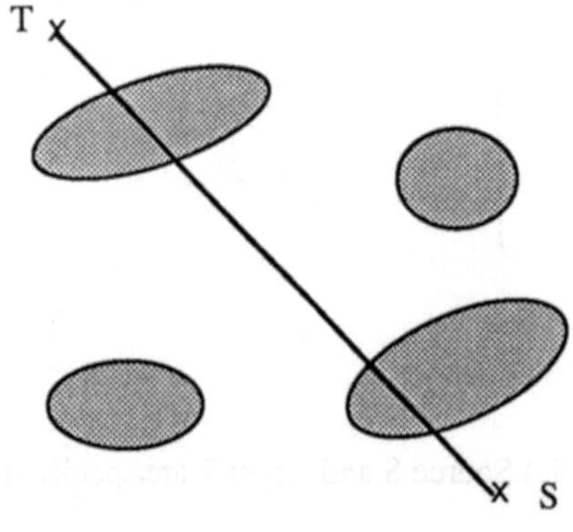

fig 6a

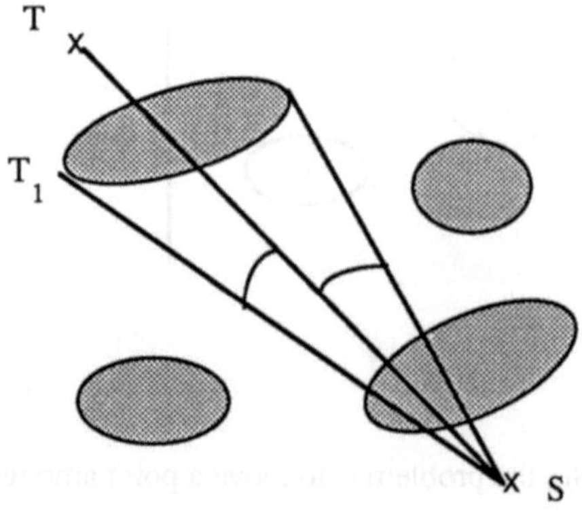

fig 6b

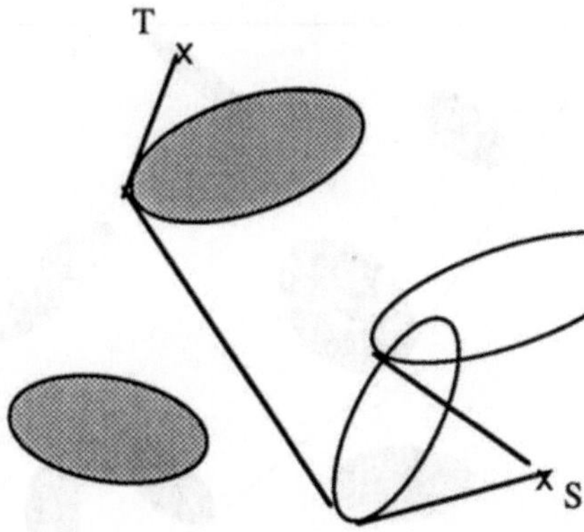

fig 7

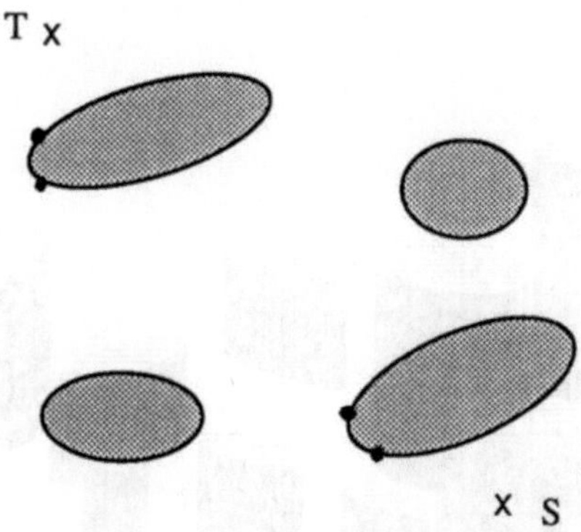

fig 8

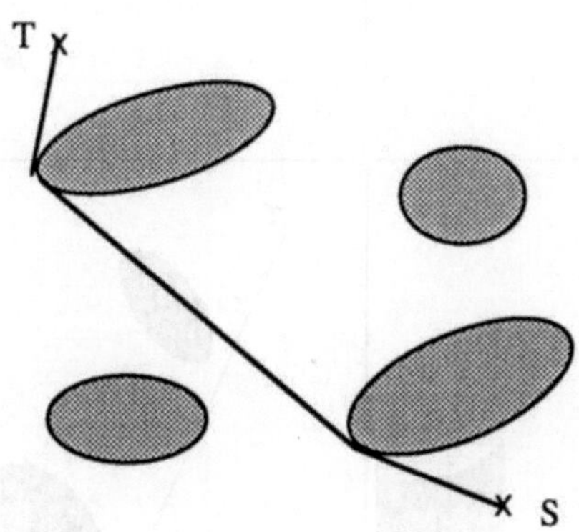

fig 9

III) IMPLEMENTATION AND SIMULATION

The algorithm is written in the "C" language and makes use of primitive functions included in a library. The simulations are performed on a SUN 3/160 using 68020 and 68881 .

The running time to get the set of points in an environment containing 10 obstacles is about 0.5 s (fig 10), but in fact we do not need to wait all this time. We can begin the movement as soon as we get the first two points. Fig 11 presents another case where grown obstacles overlap.

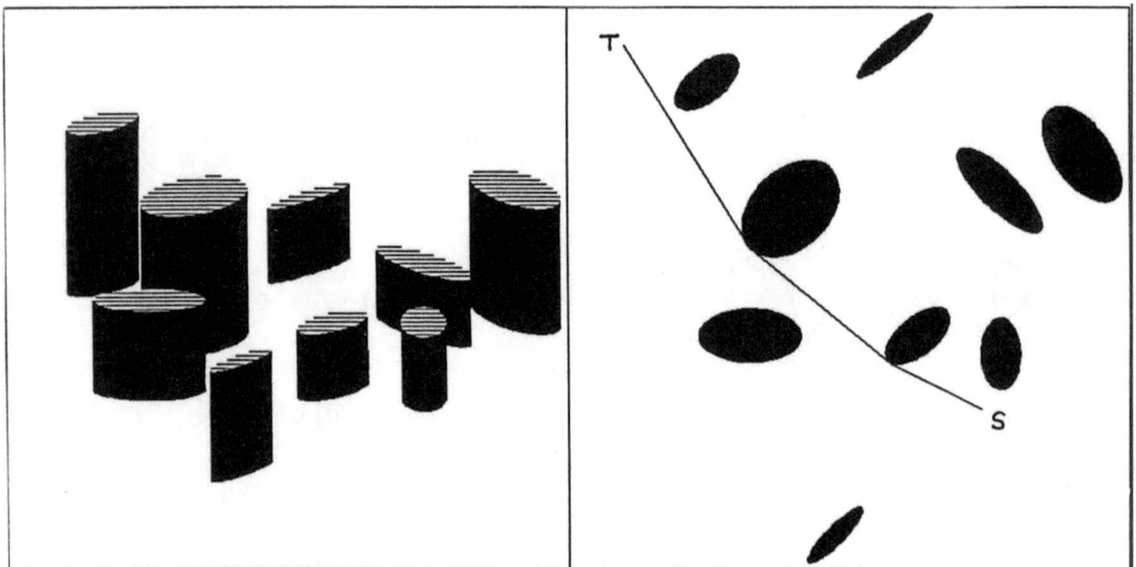

Fig. 10

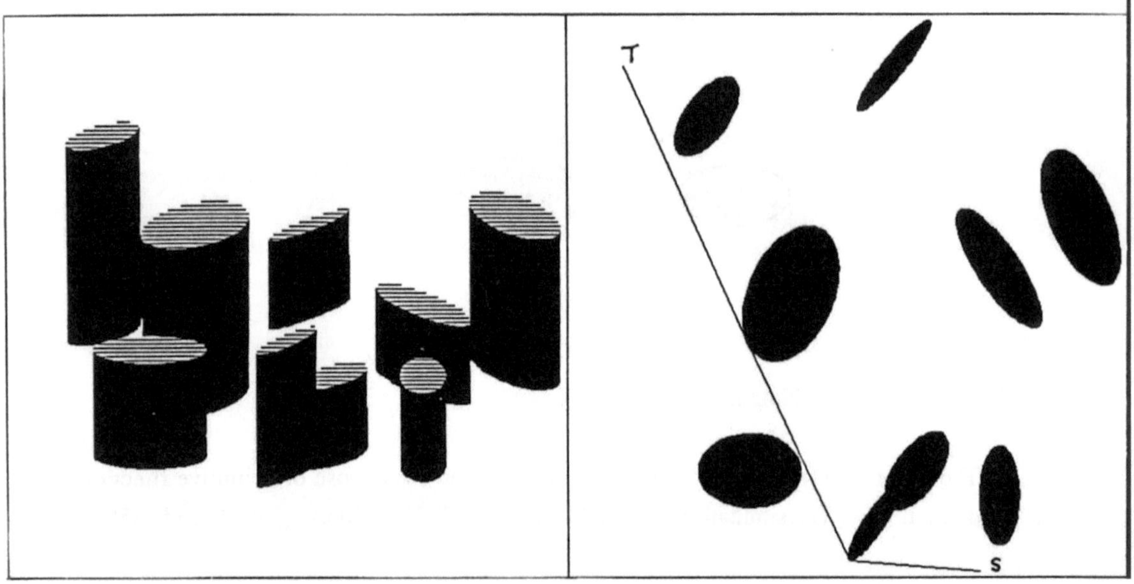

Fig. 11

IV) CONCLUSION

In this paper we have presented some requirements for a task-level language. Before developing a task-level system, we have to solve sub-problems such as : automatic grasping, collision avoidance and fine motions. We enumerate the repertoire of trajectories available to the user. Then we have considered the problem of translating an elliptic object among elliptic obstacles. A solution to the collision avoidance problem for the SCARA-type manipulator is presented. The algorithm is computationally efficient. The output of the collision avoidance system has to be interfaced with a real SCARA robot. Future extensions concern the developement of the world model, grasping planning and tools the programmer needs.

Note : The presentation of this paper was accompanied by a video tape showing recent experiments at CERT on a parallel robot.

APPENDIX

How to compute the common tangent to two ellipses :

Given equations of two ellipses E_1 and E_2

$$(E_1):\ Ax^2+By^2+Cxy+Dx+Ey+F=0 \quad (1)$$
$$(E_2):\ A'x^2+B'y^2+C'xy+D'x+E'y+F'=0 \quad (2)$$

Let the equation of the tangente be :

$$y=ix+j \quad (3)$$

Replacing y in (1) and (2) by (3), we get a second order equation in x.
(3) is the tangent only if the discriminant is null.

$$i^2(\frac{E^2}{4}-BF)+j^2(\frac{C^2}{4}-AB)+ij(BD-\frac{CE}{2})+i(\frac{DE}{2}-CF)+j(\frac{CD}{2}-AE)+\frac{D^2}{4}-AF=0$$
$$i^2(\frac{E'^2}{4}-B'F')+j^2(\frac{C'^2}{4}-A'B')+ij(B'D'-\frac{C'E'}{2})+i(\frac{D'E'}{2}-C'F')+j(\frac{C'D'}{2}-A'E')+\frac{D'^2}{4}-A'F'=0$$

To simplify the notation let us use

$$\alpha=\frac{E^2}{4}-BF \qquad \beta=\frac{C^2}{4}-AB$$
$$\gamma=BD-\frac{CE}{2} \qquad \delta=\frac{DE}{2}-CF$$
$$\theta=\frac{CD}{2}-AE \qquad \eta=\frac{D^2}{4}-AF$$

Thus

$$\alpha i^2+\beta j^2+\gamma ij+\delta i+\theta j+\eta=0 \quad F_1$$
$$\alpha' i^2+\beta' j^2+\gamma' ij+\delta' i+\theta' j+\eta'=0 \quad F_2$$

$$\frac{\partial F_1}{\partial i} = 2\alpha i+\gamma j+\delta = U$$

$$\frac{\partial F_1}{\partial j} = 2\beta j+\gamma i+\theta = w$$

$$\frac{\partial F_2}{\partial i} = 2\alpha' i+\gamma' j+\delta' = Z$$

$$\frac{\partial F_2}{\partial j} = 2\beta' j+\gamma' i+\theta' = v$$

$$J = \begin{pmatrix} \frac{\partial F_1}{\partial i} & \frac{\partial F_1}{\partial j} \\ \frac{\partial F_2}{\partial i} & \frac{\partial F_2}{\partial j} \end{pmatrix} = \begin{pmatrix} U & W \\ Z & V \end{pmatrix}$$

det = UV-WZ

$$J^{-1} = \frac{1}{\det} \begin{pmatrix} \frac{\partial F_2}{\partial j} & -\frac{\partial F_1}{\partial j} \\ -\frac{\partial F_2}{\partial i} & \frac{\partial F_1}{\partial i} \end{pmatrix}$$

Newton method

$$X = \begin{pmatrix} i \\ j \end{pmatrix} \qquad F: \begin{cases} F_1(x) = 0 \\ F_2(x) = 0 \end{cases}$$

Given Xo

$$X_{n+1} = X_n - kJ^{-1}(X_n)F(X_n)$$

Remark : Step 6 provides a good initialization.

REFERENCES & BIBLIOGRAPHY

[1] J.E. BOBROW - S. DUBOWSKY - J.S. GIBSON "Time Optimal Control of Robotic Manipulators Along Specified Paths".The International Journal of Robotics Research, Vol.4, n°3, Fall 1985

[2] F. PFEIFFER - R. JOHANNI "A concept for manipulator trajectory planning". IEEE 1986

[3] AHLBERG - NILSON - WALSH "The theory of splines and their applications"

[4] B. John OOMMEN - I. REICHSTEIN"On the problem of translating an elliptic object through a workspace of elliptic obstacles".Robotica, 1987, Vol.22, n°10.

[5] T .LOZANO-PEREZ ,M.A WESLEY "An Algorithm for planning collision free paths among polyhedral obstacles" Communications of the ACM October 1979 Vol 22 No 10 .

[6] L.GOUZENES "Strategies for solving collision free trajectory problems for mobile manipulator robots" The International Journal of Robotics Research Vol 3 No 4 Winter 1984.

[7] S.M UDUPA " Collision detection and avoidance in computer controlled manipulators" Proceedings of the IJCAI-5,MIT,Cambridge MA (August 1977).

[8] S.S YIENGAR et al ."Learned Navigation Paths for a robot in Unexplored Terrain" IEEE 1985 Second Conference on Artificial Intelligence Applications.

[9] B.FAVERJON "Obstacle Avoidance using an octree in the configuration space of a manipulator" IEEE 1984 First conference on Artificial Intelligence Applications.

[10] O.KHATIB "Real time obstacle avoidance for manipulators and mobile robots" The International Journal of Robotics Research Vol 5 No 1 Spring 1986.

[11] C.REBOULET and A.ROBERT "Hybrid control of a manipulator equiped with an active compliant wrist" 3rd ISRR Robotics Research, Cambridge, Mass, MIT Press (1985)

Fig. 12

PLANNING AND SENSING TRADEOFFS IN ROBOTICS

Mohamed S. Kamel
Paul Kaufmann
Pattern Analysis and Machine Intelligence Group
Department of Systems Design Engineering
University of Waterloo
Waterloo, Ontario, Canada, N2L 3G1

Abstract

In this paper we discuss planning and sensing in the robotic environment. Although sensing is more related to the execution of robot actions rather than their generation, this paper shows the relationship between sensing and planning. A strategy for planning and sensing tradeoff analysis is proposed. The strategy is explained in a framework of a planner that is based on a game theoretic approach. This planning approach provides the mechanism for realizing the tradeoff strategy and leads to achieving adaptable planning.

Introduction

Planning in robotics refers to the process of generating partial or complete plans for a robot to achieve a given task. A plan constitutes a high level program that the robot could execute. Therefore, planning would obviate the need for re-programming when the task changes and/or maintaining a large database of stored plans.

Sensing is required to guide the robot in the environment. It allows more flexible and less stringent control. Although sensing is more related to the execution of robot actions rather than the generation of these actions, it can have a significant impact on the planning approach used to generate these actions.

Traditionally in planning, the environment is modelled and all required information is assumed to be known a-priori. Models representing the environment are usually ideal, deterministic and totally predictive. Side-effects and interactions cannot be represented and the environment is assumed to only change through robot actions. Obviously the real world is not like that, hence better models of the environment are necessary for the success of planners in solving real world problems. The level of planning will depend on the predictive power of the model. Recently, researchers have suggested "reactive" planning which takes the extreme of using no prediction at all. Plans in this approach are generated during execution according to current situations. Such planners can rely heavily on sensing to provide information on the current state of the environment. Although this approach overcomes the modelling problem, it reduces

NATO ASI Series, Vol. F 64
Sensory Robotics for the Handling of Limp Materials
Edited by P. M. Taylor

the prediction and in turn the planning power.

What is needed are approaches that are more adaptable ; approaches that provide a balance between sensing and planning by interleaving between generating the plan and executing it. In such approaches, a tradeoff analysis between planning and sensing is required.

In this paper, we discuss a strategy for planning and sensing tradeoff analysis in a framework of a planner that is based on a game theoretic approach.

Early Work in Planning

To date, there have been several notable contributions to planning. The General Problem Solver, or GPS, [ERNST69] introduced the concept of a means-ends analysis. This refers to using the difference between the current state and the goal state to select operators. STRIPS, a planner designed by Fikes et. al. [FIKES71], uses this notion for solving simple robot task problems. STRIPS demonstrated the use of the first-order predicate calculus and automatic theorem proving in problem solving. Later, a system for storing and generalizing plans was introduced [FIKES72].

Sacerdoti [SACER74], investigated the idea of planning in a hierarchy of abstract spaces. This refers to the developing of a plan in increasing levels of detail, identifying and addressing the most critical subgoals (a sub component of the goal) ahead of the less critical. His design, based on STRIPS, was called ABSTRIPS. Later, Sacerdoti developed an improved planner known as NOAH [SACER75]. NOAH is also a hierarchical planner which introduces the concepts of a *procedural net* and planning *critics*. A procedural net represents several levels of a plan. At the top is the most abstract goal. Each subsequent level consists of *partially ordered* sequences of nodes (the order of some nodes is fixed) representing subgoals. At the lowest level, each subgoal is a primitive (executable) planning operator. Critics are programs used by the planner to watch for and correct effects of actions that may jeopardize the success of the plan.

One important problem in planning is the treatment of interacting subgoals. Subgoal interaction may occur when the goal involves the attainment of non-independent conjunctive subgoals. Planners that attempted addressing this issue are HACKER [SUSSM73], INTERPLAN [TATE75] and a system by Waldlinger [WALDI77]. The planner NOAH also deals with interacting subgoals through the use of critics. As a plan is developed, subgoals are left unordered (it is assumed they can be done in parallel) until it becomes apparent that ordering is necessary. A critic notes interactions between operations and corrects these by ordering them. Sacerdoti's ideas of postponing the ordering has been developed further in the SIPE planner [WILKI84].

The planners mentioned above are alike in that they assume that all required information is known a-priori, all operators have known unique outcomes, side-effects and interactions cannot be represented, and the environment is unchanging (unless through robot actions). As humans, we know that very few decisions are made under such ideal conditions. In many application domains uncertainty is introduced by the robot actions, the sensors and material being handled (Limp material is a good example). Uncertainty could also exist due to incomplete specification of the environment.

Planners designed for operation under uncertain conditions are few.

Munson [MUNSO77] presents an excellent formal discussion on deterministic planning and extends this to planning under uncertainty.

Munson points out that game theory is the "natural vehicle" for planning in uncertain environments. However, he is pessimistic as to its value owing to "our primitive capabilities in quantifying real environments". Certainly any probabilistic planner faces the same problem. The question is to what extent must real world events be quantified for the planning system to be a useful decision making tool?

Feldman [FELDM77] discusses in detail the role decision analysis can play in symbolic problem solving (of which planning is an example). He uses the monkey and bananas problem to illustrate several important decision theoretic concepts such as the utility function (used to measure the worth of a state under conditions of uncertainty), the goal of maximizing utility, and the correct treatment of operators used to gather information (ie.: perform a test in the environment). He focuses on using the utility measure of a skeleton plan to guide plan elaboration and the decision of when to execute versus when to plan.

The work of Farley [FARLE83] represents, in the authors' opinion, the only serious attempt at actually describing a specific system for uncertain problem solving. Farley defines an uncertain state as a set of states each having a fixed probability of occurring. An unreliable operator is modeled by a precondition-effect pair with each effect having a stated probability. The application of an unreliable operator to an uncertain state results in a new uncertain state that combines all the effects, specified by the operator, for those preconditions that were met.

Farley's work represents an important contribution to uncertain planning. However, its representation of operators and states is not sufficiently powerful to model complex robot task environments. Operators require explicit mappings between state characteristics and multiple outcomes which have fixed a-priori probabilities of occurring. In addition, there is no way of linking effects to models of probabilistic events that could take place in the environment (e.g. the failure of a component).

It is observed from the above that traditional deterministic planners require complete knowledge and total predictability of the environment and the planning is done in an offline setting (independent of the execution). Dealing with uncertain environments has motivated the need for 1) nondeterministic models of the environment and 2) interleaving planning with information gathering operations when the models cannot predict accurately the state of the environment.

Recent Advances

In recent years, new approaches have been introduced to solve the planning problem. Among these approaches is the case-based planning approach [HAMMO86]. In this approach, which has its roots in Sussman's Planner (HACKER, [SUSSM73]), the emphasis is on learning from debugging plans and once a good plan is arrived at it is stored for later retrieval to solve similar problems. Different techniques for organizing the learned knowledge have been proposed [LAIRD87], [SIMMO87].

Another approach that has been supported by some researchers from MIT, is what is called "reactive planning". In this approach the planner reacts to current situations, no prediction about the environments is done and plans are generated during execution [CHAPM86], [FIRBY87]. The claim is that reactive planners will be more successful especially when the environment is not well specified and information may be acquired through the performance of some actions. Other related approaches have been proposed to deal with uncertain environments [WILKI85],[BROOK85], [ROSEN87]. While traditional planners assume complete knowledge of the environment through their models, reactive planners go to the other extreme and assume no prediction of the environment. In the case of uncertain environments both approaches will have problems. Accurate models of such environments do not exist and relying totally on reactions does not provide the planning power and choice of correct actions. Approaches which provide a balance between the two extremes may be more suitable for such environments. The authors have developed Probabilistic representation schemes for representing uncertainty in the robot operators and the environment as an attempt to provide better models for uncertain environments [KAMEL88]. These schemes are to be used with a probabilistic planners that provides a compromise between totally predictive and purely reactive planners [KAUFM87]. The game theoretic approach we will dwell upon in this paper is the basis of such planner, however, here we are more concerned with the issue of when to sense and when to plan. The results are applicable to probabilistic as well as non-probabilistic planners.

The Game Theoretic Approach to Planning

From the artificial intelligence perspective, game playing is a search problem in which two opponents alternately execute moves thus changing the state of the game. Each player's goal is to win the game and so the moves that one makes tend to impede the other.

As in other search problems, the search space is explored by expanding a search tree, in this case a game tree. The tree is expanded from one player's perspective in an effort to decide his best next move. The expansion is done up to a specified depth and the leaf nodes are then rated. The rating is a measure of the proximity to a win and is generated by an *evaluation function* . This heuristic device is necessary since time and storage requirements will usually prohibit complete expansion of a game tree (to the point of finding a leaf state representing a win).

After evaluating leaf states, the evaluation function values may be backed up in order to provide information for the player to choose among the alternatives for the next move. Among algorithms for performing this backup is the MINIMAX algorithm (see for example [WINST77]). The MINIMAX algorithm involves alternately backing up maximum values of successor ratings to the player and minimum values to the opponent. The algorithm assumes that each player will always endeavor to make his best move. The player makes a move then awaits the opponents reaction and the process for planning the next move repeats. This is interleaving between planning and sensing.

Thus the game playing paradigm seems ideal for the type of planning we want. In this scenario, the player is the robot whose task is to achieve the specified goal. One can imagine a 'demon' in control of *the player's view of the environment* as an adversary. The moves of the adversary represent possible reactions of the environment to actions taken by the player.

The main difference between game playing and planning is the predictability of the opponent's moves. In game playing the opponent is assumed to be adverse and wanting to win the game. In planning the environment's reactions don't necessarily mean worst responses to the robot actions. The reactions may be helpful or of no effect and in some cases they may be harmful. In [KAMEL88], we have proposed a representation for the robot actions and the reaction operators. Robot actions are defined as macro-operators, or M-OP's. A M-OP is an operator used at a high level of abstraction as it does not give consideration to low level details of its execution. For example, a M-OP used to move an object, would ignore details of gripper orientation and trajectory. Indeed, if necessary, a version of the planner could be used at a lower level of abstraction to form a plan for the M-OP's execution.

A M-OP can be either manipulative if it performs some physical action in the world, sensory if it passively observes and records some aspect of the environment, or acquisitional if it tests some aspect of the environment to acquire new information. The latter combines sensory and physical operations. The discussion here will be confined to manipulative M-OP's. Furthermore, in keeping with the view that an adversary is responsible for generating multiple outcomes, M-OP's are modeled as reliable, that is having a unique outcome. The moves of the adversary are called R-OP's (reaction operators).

Notice that we have not focussed on a particular kind of robot action nor presupposed that some outcomes are more desirable than others (although this may be the case). Indeed, any uncertainty related to the execution of any operator will be manifested as that operator having several possible outcomes.

Besides being an elegant conceptualization, there are distinct advantages to modeling the reactions of the world separately from the macro-operations that precipitate them. Firstly, a planning system designed for a particular robot should be as flexible as possible. It will likely be necessary to fix the world and M-OP representations. However, there remains a great deal of flexibility if the (human) user of the system can add or subtract R-operator definitions as required by the robot's current workspace and/or by its changing performance. Secondly, sequences of R-OP's could be used as building blocks to generate more complex states than could be generated by one operator. Thirdly, it is possible that several M-OP's may, under conditions of uncertainty, precipitate similar multiple outcomes. If multiple outcome generation is represented on the M-OP definition then there could be a great deal of duplication. As an example, consider the R-OP (KNOCK_LEFT O_i) which models knocking an object O_i to the left in an attempt to grip it. The re-positioning of the object is an outcome generated by the R-OP [KAMEL88].

Another difference between game playing and planning is that in game playing (most games), the opponent's turn is predefined and an update of the current situation is bound to happen. In planning as in game playing it is desired to have this update only when required. Measures that could be used to trigger the need for the update need to be defined. These may include measures of the level of uncertainty in the environment, the utility of the update, occurrence of reactions and/or detection of failure or error.

The game theoretic model provides a good framework upon which to base adaptable task planners. However, its benefits go beyond the conceptual level. Apart from the advantages of modelling uncertain events separately (i.e., as in an adversary's moves) from the robot operations that precipitate them, it enables the application of techniques from game and decision theory to the planning problem, if not directly then with little modification.

A Strategy for Planning and Sensing Tradeoff

A planner based on the approach discussed above calls for a mechanism to decide when to plan, when to execute and when to update (sense) the environment. Tradeoff analyses need to be performed in order to make these decisions. Planning time, update time, update utility, uncertainty, and the type of problem are some of the factors that could be involved in this analysis.

Uncertainty is measured in two ways. First, the probability of the occurrence of reaction operators is estimated. This gives a measure for changes in the environment as a result of the robot actions. Second, the level of uncertainty in the states that result from M-OP and R-OP is evaluated.

Failure and the Unknown Response

Let us suppose that each outcome is generated by one R-OP. Let r_i denote an event that occurs during M-OP execution that could lead to an outcome represented by some R-OP. This event arises due to some uncertainty in the world representation (for example, gripper uncertainty). In this section, an approach to computing the values of $P(r_i)$, the probability that the R-OP event will occur is presented. Also, the possibility of an unknown response (P(ur)) occurring must be considered. This is an important issue which will also be discussed.

Together, all responses must satisfy the following equation:

$$1 = P(\bar{r}) + \sum_{i=1}^{n} P(r_i) + P(ur) \qquad (1)$$

where n is the number of responses and $P(\bar{r})$ is the probability of no response. The probability of a M-OP being successfully executed will be at least $P(\bar{r})$.

The probability of an R-OP event depends on the state S that exists at the time of M-OP application. The planner partitions S into different states as a result of applying test procedures to check whether the preconditions (PC) of the M-OP are satisfied. One state, say S_1, will be the state in which all M-OP PC's are met. M-OP usage in the remaining states would yield undesirable effects. These effects should be accounted for in some way. This accounting could be described by R-OP events given a powerful enough model (i.e., good R-OP definitions including recursion, simultaneous R-OP events). However, there will always be a limit to the models predictive power. Beyond this limit it cannot be said what events may take place. This is the unknown response (ur) leading to some unknown state. Quantification of P(ur) is especially important when the model is weak but will be necessary in general. For example, it is difficult for a model to predict the state resulting from a roving robot smashing into a table and toppling objects on its surface, and yet, it is necessary to know when something like this could have occurred and with what probability. The general approach to computing the terms in equation (1), assuming that the planner is investigating M-OP M in state S, would be to look at all partitioning of S and allow R-OP responses to act in response to M applied to each partition. This is illustrated in Figure 1.

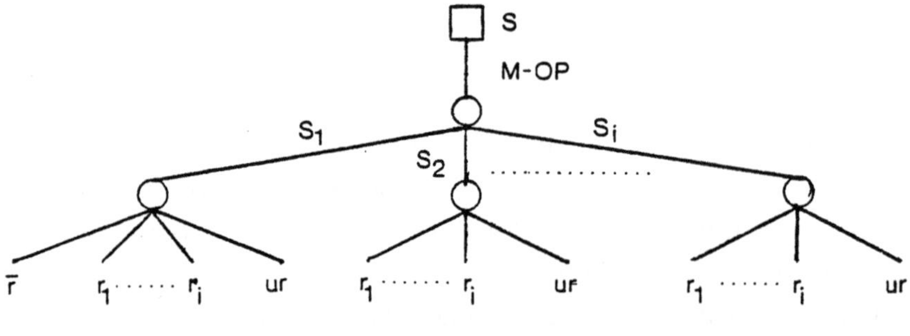

Figure 1: Responses to a M-OP applied in an uncertain state.

In this diagram S_1 represents the partitioning in which all M-OP PC's are met. Notice that $\bar{r}$ is absent below partitioning other than S_1. This is because some R-OP event will always occur if a M-OP is executed when any of its PC's are not met. From Figure 1, it can be seen that the terms in equation (1) will be:

$$P(ur) = \sum_{q=1}^{m} P(ur|S_q)P(S_q) \tag{2}$$

$$P(r_i) = \sum_{j=1}^{m} P(r_i|S_q)P(S_q) \tag{3}$$

$$P(\bar{r}) = 1 - P(ur) - \sum_{i=1}^{n} P(r_i) \tag{4}$$

where m is the number of partitioning and n the number of possible R-OP events.

Currently, multiple outcomes have been considered only for M applied to a state in which all PC's are met. One could consider this to be the limit of the current model. Accordingly, $P(r_i|S_q)=0$ for q>2, $P(ur|S_q)=1.0$ for q>2 and so P(ur) must be at least 1-P(S_1). Equations (2) and (3) become:

$$P(ur) = P(ur|S_1)P(S_1) + 1 - P(S_1) \tag{5}$$

$$P(r_i) = P(r_i|S_1)P(S_1) \tag{6}$$

As an example, suppose M involves gripping some object. Let r_1=gripper off to left and r_2=gripper off to right. The conditional terms in equations (5) and (6) could be pre-set or determined from the known gripper uncertainty. Figure 2 illustrates the latter for a gripper uncertainty represented by a standard normal distribution with standard deviation σ .

The above scenario accounts for events that occur (or do not occur) at a specific stage during M-OP execution. In general, there may be several such stages at which we would like to link some real uncertainty to R-OP events. These stages refer to groups of responses, defined by the same distribution, that become valid at some point during M-OP execution. These probability measures can be used in the tradeoff analysis to decide whether the environment has changed in unknown ways or in a manageable way. If the probability of unknown responses turns out to be higher than a specific threshold, sensing may be wise to update the environment and reduce the uncertainty and in turn the planning complexity.

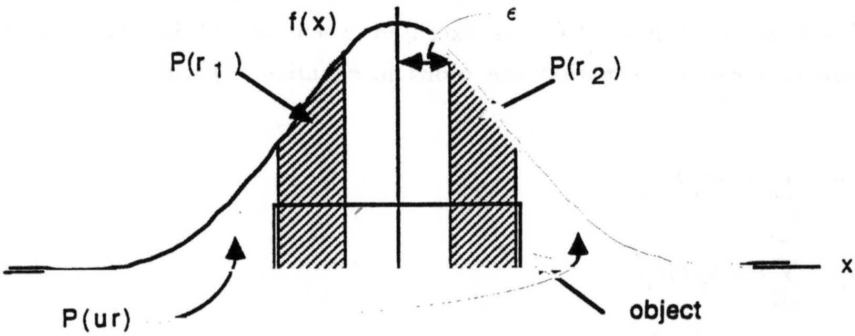

Figure 2: R-OP event probabilities from a known uncertainty

State Uncertainty and Update Cost

The level of uncertainty in a state S is indicative of how successful operations on S will be. If the uncertainty in S is high, above some threshold for example, then no action can be performed and an UPDATE operation should be performed instead. Below the threshold value, one may hypothesize that the higher the current uncertainty the more likely an UPDATE operation will be at some future time. In each case, the level of uncertainty can be expressed as a cost of updating (information gathering in general). The level of uncertainty in S will be measured using Shannon's entropy:

$$entropy(S) = -\sum_{i \in R(S)} f_i(S) \log(f_i(S)) \tag{7}$$

Where, $f_i(S)$ is the probability distribution for state S. Let $C_u(S)$ be the cost due to state uncertainty. Let ET be the entropy threshold discussed above and $C_{op}(UPDATE)$ be the update cost. Then $C_u(S)$ is expressed by:

$$C_u(S) = \begin{cases} C_{op}(UPDATE) & entropy(S) \geq ET \\ C_{fu}(S) & entropy(S) < ET \end{cases} \tag{8}$$

where C_{fu} estimates the cost of future updating based on the current entropy in S. The cost of future updating could be a proportion of the update cost. For example C_{fu} could be taken as K * $C_{op}(UPDATE)$, with K=entropy(S)/ET.

Update cost could be calculated from the sensing parameters and configurations. The techniques developed in [TAYLO88] may be useful in providing these calculations.

The worth of a state will be measured in terms of a cost function defined below:

$$C_{plan}(S,S_g) = C_{pp}(S) + C(S,S_g) \tag{9}$$

$C_{pp}(S)$ is the cost of the partial plan up to state S and is the accumulated cost of M-OP's between the root and S. $C(S,S_g)$ is the estimated cost of getting from S to the goal. $C(S,S_g)$ consists of two components: $C_t(S,S_g)$, the estimated cost of transforming S into the goal and $C_u(S)$, the level of uncertainty in S.

In computing C_t a few factors must be considered. First, if S is not a goal state but is unexpandable then S is a dead-end and C_t is set to $+\infty$. Secondly, if S is unknown it will have to be updated so $C_u(S)=C_{op}(UPDATE)$. In addition, if a state is unknown nothing can be said about its proximity to the goal and C_t has to be set to a large value. For all other cases C_t is taken as the minimum sum of operator cost over all the alternative sequences of operators that are applicable to the current state.

The cost of the plan can be used for tradeoff analysis involving response time, or a tradeoff between update cost and planning cost. Other situations where update may occur include: high likelihood that a subgoal has been met or when it cannot decide what outcome of an operation occurred. Updating will also occur when there are reasons to believe that the world is in the goal state. This guarantees that the goal will be met provided it is within the robot's ability.

Conclusions

The interleaving between planning and sensing is important for planning in environments that are complex, dynamic or uncertain. The game theoretic approach to planning provides a good framework for planners targeted to these environments. The approach allows for explicit tradeoff analyses which lead to achieving adaptable planning. Measures for uncertainty, planning cost and action imprecision have been developed and their use in the tradeoff analysis has been discussed.

References

BROOK85 Brooks, R. A. *A Robust Layered Control System for a Mobile Robot*, Tech. Report AI Memo 864, AI Lab., MIT, 1985.

CHAPM86 Chapman. D. and Agre, P. E., *Abstract Reasoning as Emergent from Concrete Activity*, In Workshop on Planning and Reasoning about Action, Portland, Oregon, 1986.

ERNST69 Ernst, G. and Newell, A., *GPS: A Case Study in Generality and Problem Solving*, Academic Press, New York, 1969

FARLE83 Farley, A. M., *A Probabilistic Model for Uncertain Problem Solving*, IEEE Trans. on SMC, vol. SMC-13, no. 4, pp.568-579, 1983.

FIKES71 Fikes, R. E. and Nilsson, N. J., *A New Approach to the Application of Theorem Proving to Problem Solving*, Artificial Intelligence, 2, pp.189-208, 1971.

FIKES72 Fikes, R. E., Hart, P. E. and Nilsson, N. J., *Some New Directions in Problem Solving*, Machine Intelligence, vol. 3, pp.405-430, 1972.

FIRBY87 Firby, R. J., *An Investigation into Reactive Planning in Complex Domains*, Proc. of Sixth National Conference on Artificial Intelligence, AAAI, Seattle, Washington, pp.202-206, 1987.

HAMMO86 Hammond, K., *CHEF: A Model of Case-Based Planning*, Proc. of Fifth National Conference on Artificial Intelligence, AAAI, Menlo Park Calif., pp.267-271, 1986.

KAMEL88 Kamel, M. and Kaufmann, P., *Representing Uncertainty in Robot Task Planning*, Proc. of IEEE International Conf. on Robotics and Automation, Philadelphia, pp.1728-1734, 1988.

KAUFM87 Kaufmann, P., *Robotic Task Planning in Uncertain Environments*, Masters Thesis, Department of Systems Design Engineering, University of Waterloo, 1987.

LARID87 Laird, J.E., Newell, A. and Rosenbloom, P.S., *SOAR: An Architecture for General Intelligence*, Artificial Intelligence, 33, pp.1-64, 1987.

MUNSO71 Munson, J. H., *Robot Planning, Execution and Monitoring in an Uncertain Environment*, Proc. Second IJCAI, p.338-349, Sept., 1971.

ROSEN87 Rosenschein, S., Kaelbling, J. and Pack, L., *The Synthesis of Digital Machines with Provable Epistemic Properties*, In Theoretical Aspects of Reasoning about Knowledge, Proc. of 1986 Conf., ed. J. Halpern, Los Altos, Calif., Morgan Kaufmann, 1987.

SACER74 Sacerdoti, E. D. *Planning in a Hierarchy of Abstract Space*, Artif. Intel., 5:115-135, 1974.

SACER75 Sacerdoti, E. D., *A Structure for Plans and Behavior*, Tech. Note 109, AI Center, SRI International, 1975.

SIMMO87 Simmon, R. and Davis, R., *Generate, Test, and Debug: Combining Associational Rules and Causal Models*, In Proc. of the Tenth International Conf. on Artificial Intelligence, AAAI, Los Altos Calif.,pp.1071-1078, 1987.

SUSSM73 Sussman, G. J. *A Computational Model of Skill Acquisition*, AI Tech. Report 297, AI Laboratory, MIT (Doctoral Dissertation), 1973.

TATE75 Tate, A., *Interacting Goals and their Use*, IJCAI 4, pp.215-218, 1975.

TAYLO88 Taylor, G.E. and Taylor, P.M., *Dynamic Error Probability Vectors: A Framework for Sensory Decision Making*, Proc. of IEEE International Conf. on Robotics and Automation, Philadelphia, pp.1096-1100, 1988.

WALDI77 Waldinger, R., *Achieving Several Goals Simultaneously*, In E.W. Elcock and D. Michie (eds.) Machine Intelligence, Halstead/Wiley, 1975.

WILKI84 Wilkins, D. E., *Domain-independent Planning: Representation and Plan Generation*, Artificial Intelligence, 22, pp.269-301, 1984.

WILKI85 Wilkins, D. E., *Recovering from Execution Errors in SIPE*, Computational Intelligence, vol. 1, no.1, 1985.

ANALYSIS OF A PROBABILISTIC FRAMEWORK FOR AN INTELLIGENT WORKCELL

P M Taylor, G E Taylor, I Halleron, X K Song

The Department of Electronic Engineering
The University of Hull
Hull. HU6 7RX. UK

ABSTRACT

This paper describes work done at Hull towards answers to the question, "what sensing actions should be taken, by what sensors, when, and how should the information be acted upon"?, raised in the development of an intelligent workcell. The work provides an analysis of the probabilistic dynamic error vector framework described in [1]. A ply separation and motif application task is then used to illustrate the operation of the framework.

Keywords: robot, automation, garment assembly, error recovery, probabilistic.

INTRODUCTION

A robotic workcell for use in a batch production environment may be equipped with many sensors to provide it with tolerance to uncertainties in components, actuators, fixtures and the sensors themselves. To use all these sensors throughout the production process would be costly, and so raises the question posed in [1,2], "what sensing actions should be taken, by what sensors, when, and how should the information be acted upon"? The design of sensory assembly systems, and the process of ensuring that they work correctly is

NATO ASI Series, Vol. F 64
Sensory Robotics for the Handling of Limp Materials
Edited by P. M. Taylor

time consuming and costly and so work at Hull is aimed at the automation of this process, thus reducing small batch production costs.

Figure 1. shows how an offline cell controller containing a knowledge base with information about assembly operations, sensors, actuators, cell errors and error recovery is supplied with an object level description of the assembly task undertaken. Based on the probabilities of possible workcell errors, a manipulator level task description is produced which includes the sensing strategy and error recovery actions which will be used in the assembly task. This task description will then be downloaded to the online cell controller which will modify its actions in accordance with the sensory information and improved knowledge of error probabilities obtained during run time. The runtime experience can be used periodically to update the offline knowledge base.

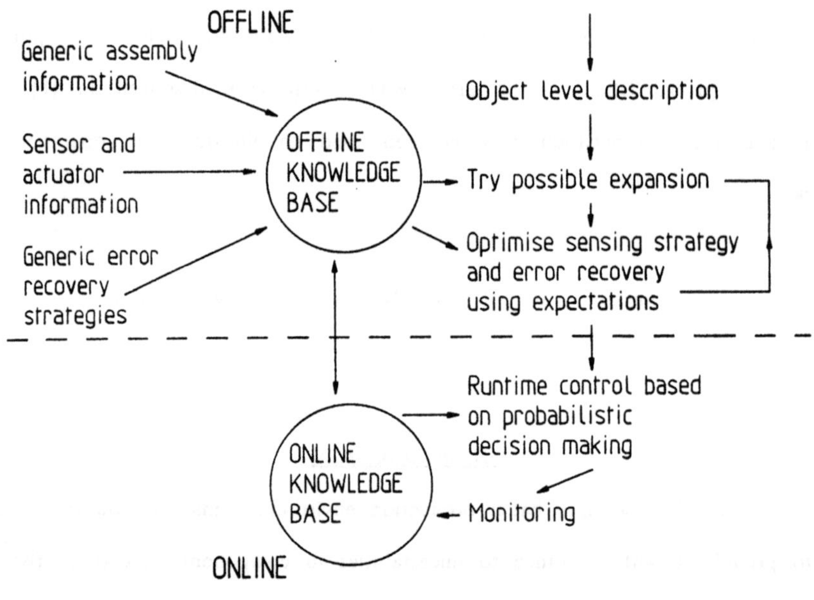

Figure 1. Autonomous workcell:
offline design, online execution.

The following describes a probabilistic decision making process to be used in both the offline and online cell controllers. Offline, the aim is the optimal selection of a sensor set for each stage in the assembly task, which will economically detect a subset of all possible error causes at that and other stages. Economical detection should result in the minimisation of the overall cost of producing a batch of the product. Online, the decision algorithm is used for the diagnosis of error causes from the available sensory information.

This paper commences with the definitions of the vectors and matrices used in the decision making process, followed by the definition of a lower bound, an upper bound and a best guess operator used to manipulate the vectors and matrices in the 'forward' calcualation, examples being given in each case. The use of the inverse, †, operator and Bayes theorem in the 'backwards' calculation are then described using suitable examples.

ERROR PROBABILITY VECTORS

The relationship between error causes and cell sensors is formalised as follows.
The error *causes* (vacuum supply failure, damaged object, ...) have related error *effects* on the object (no object on gripper, damaged object on gripper, ...), which may be detected by the measurement of some *parameter* (vacuum incorrect, zero weight, non-zero but incorrect weight, incorrect object area, ...). A selected set of *sensors* (air pressure sensor, load cell on gripper, camera, ...) are used to determine these parameters. Thus we have four vectors:

error cause vector c

error effect vector e

sense parameter vector p

sensor vector s

which are related in the following sequential manner,

$$c \Longrightarrow e \Longrightarrow p \Longrightarrow s$$

The *error cause vector*, c, has n entries, where n is the number of possible error causes at a given assembly stage, and the values of the entries in c represent the probabilities of each of these causes. Thus

$$c = (c(1), c(2), \dots , c(n))^T \qquad 0 \le c(i) \le 1 \qquad \forall i$$

$$\sum_i c(i) = 1$$

c_n represents the ideal case, a null error. The sum of the error causes being unity implies they are mutually exclusive and exhaustive. In a real robot cell the error causes may not be mutually exclusive, but if two (or more) errors were to occur simultaneously then they could be recovered from sequentially. Correlated errors are likely to be different manifestations of a single cause.

The *error effect vector*, ***e***, contains m entries corresponding to the number of error effects implied by ***c***. The elements in ***e*** give the probabilities of the various effects.

$$\boldsymbol{e} = (e(1),\ e(2),\ \dots,\ e(\mathrm{m}))^{\mathrm{T}} \qquad 0 \leq e(\mathrm{i}) \leq 1 \quad \forall \mathrm{i}$$

The effect vector may be smaller than the cause vector (a number of causes produce the same effect) or larger (a cause produces several distinct effects).

The *sense parameter vector*, ***p***, contains k entries corresponding to the parameters which the errors may affect. The entries in ***p*** give the probabilities of the various parameters being at error.

$$\boldsymbol{p} = (p(1),\ p(2),\ \dots,\ p(\mathrm{k}))^{\mathrm{T}} \qquad 0 \leq p(\mathrm{i}) \leq 1 \quad \forall \mathrm{i}$$

The *sensor vector* ***s*** contains q entries corresponding to the sensors in use. The entries in ***s*** give the probability of a sensor (or sensory process) indicating an error.

$$\boldsymbol{s} = (s(1),\ s(2),\ \dots,\ s(\mathrm{q}))^{\mathrm{T}} \qquad 0 \leq s(\mathrm{i}) \leq 1 \quad \forall \mathrm{i}$$

The above vectors need not be of the same dimension, as explained for the cause and effect.

TRANSFORMATION MATRICES

The vectors ***c, e, p, s*** are related by the three matrices ***C, E, P*** as

$$\boldsymbol{e} = \boldsymbol{C} * \boldsymbol{c}$$

$$\boldsymbol{p} = \boldsymbol{E} * \boldsymbol{e}$$

$$\boldsymbol{s} = \boldsymbol{P} * \boldsymbol{p}$$

The nature of the operator * will be discussed shortly. C is a constant m x n matrix whose elements $c(i,j)$ represent the conditional probability that cause j is manifested by effect i. That is $c(i,j) = p(e(i) \mid c(j))$, $c(i,j) \leq 1, \forall i,j$. Similarly the matrix E is a k x m matrix of the form $e(i,j) = p(p(i) \mid e(j))$, describing the relations between the error effects and the parameters affected, and P is a q x k matrix with elements $p(i,j) = p(s(i) \mid p(j))$ which describe probabilistically the sensors which are affected by each of the parameters. Thus we also have $e(i,j), p(i,j) \leq 1, \forall i,j$. Often the elements of these matrices will be unity, for example an empty feeder will always result in 'no object on gripper'. There will be many cases, however, when the elements are not unity; a misplaced bin of components may not always result in failure to grasp an object. Noise may play a rôle here, affecting, for example, the weights of objects. Also the sensors may not be totally reliable.

<u>Matrix combination</u>

Using the chain rule we can say that

$$\boldsymbol{s} = p(\boldsymbol{s} \mid \boldsymbol{p}).p(\boldsymbol{p} \mid \boldsymbol{e}).p(\boldsymbol{e} \mid \boldsymbol{c}).p(\boldsymbol{c})$$

$$\boldsymbol{s} = P*E*C.c$$

If the elements of $\boldsymbol{e}$ and $\boldsymbol{p}$ were mutually exclusive then * would represent ordinary matrix multiplication. However, in most cases the elements are not mutually exclusive and so new operators have to be defined.

A development of this formalism allows the 'no error' state (O.K. state) to be represented in the vectors and matrices so that the cell controller uses the matrix system to select its next action from either: a set of recovery sequences, or, the next step in the assembly routine.

These matrices are used for two purposes. The <u>*forward calculation*</u> is of the form

$c \longrightarrow \boldsymbol{e} \longrightarrow p \longrightarrow \boldsymbol{s}$ and obtains the $\boldsymbol{s}$ vector from the c vector in order to:

a). see which sensors are sensitive to the various error causes. c vectors representing a single error cause, in the form (0 1 0 0 0 0 0), are used for this.

b). produce a single transformation matrix of the form $T = P*E*C$ which can be used in the backward calculation.

The ***backward calculation*** is carried out to deduce the error cause responsible for the current sensor information s. This is primarily a Bayesian calculation using the T matrix and an a priori c vector containing the estimated probabilities of the various error causes before the sensor information was available. A similar calculation may be done to deduce the effects responsible for s.

Paths

In the combination of matrices we may consider 'paths'. An example explains what this means. In the combination of the E and C matrices to form the T matrix we might form paths linking the elements c_1 and p_1 via the three elements of e, as shown in figure 2.

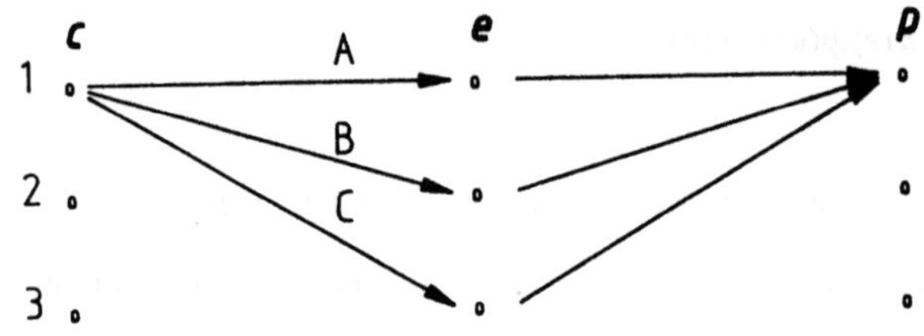

Figure 2. Path format.

Path A, $p(A) = p(p_1 | e_1)p(e_1 | c_1)$

Path B, $p(A) = p(p_1 | e_2)p(e_2 | c_1)$

Path C, $p(A) = p(p_1 | e_3)p(e_3 | c_1)$

These paths could be mutually exclusive, independent or dependent. For example we could have the case shown in figure 3.

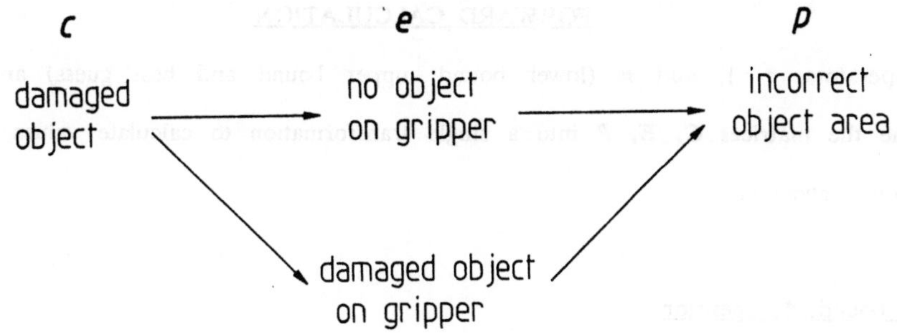

Figure 3. Example of paths.

The two paths A and B are mutually exclusive since it is not possible to have 'no object on gripper' at the same time as 'damaged object on gripper'. Similarly it is possible to have independent or dependent paths.

The probability of propagation of information (i.e. $p(p_1 | c_2)$)for a case with three paths is as follows:

$$T = p(A) + p(B) + p(C) \quad (1)$$

if the paths are mutually exclusive

$$T = p(A) + p(B) + p(C) - p(AB) - p(BC) - p(AC) + p(ABC) \quad (2)$$

where $p(XY) = p(X)p(Y)$, $\forall$ X,Y

if the paths are independent

$$T = p(A) + p(B) + p(C) - p(AB) - p(BC) - p(AC) + p(ABC) \quad (3)$$

where $p(XY) \neq p(X)p(Y)$, $\forall$ X,Y

if the paths have inter-dependencies.

We are assuming no information on the dependencies between these paths and so must evaluate upper and lower bounds and a best guess value for the matrix T, using different operators to combine the C, E and P matrices. Each of these operators is considered below.

FORWARD CALCULATION

The operators, *, !, and # (lower bound, upper bound and best guess) are used to combine the matrices ***C, E, P*** into a single transformation to calculate values of any of the vectors given ***c***.

Lower bound, *, operator

The operation * between two matrices ***X*** and ***Z*** is defined as

$$X = Y * Z \tag{4}$$

$$x(i,j) = \max \{ y(i,k).z(k,j) \} \qquad \forall k$$

producing a lower bound transformation matrix by selecting the single maximum path length from those considered. That is $T = \max\{p(A), p(B), p(C)\}$. This operator gives an accurate result if the paths are subassumptive (e.g: $C,B \subset A$) as depicted in figure 4.

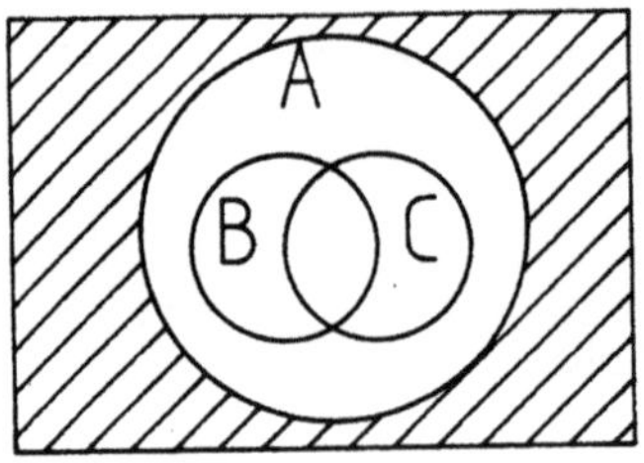

Figure 4. Paths are subassumptive.

If the paths are in fact mutually exclusive, independent or dependent to an unknown degree, as shown in figure 5,6. then this operation produces a lower bound.

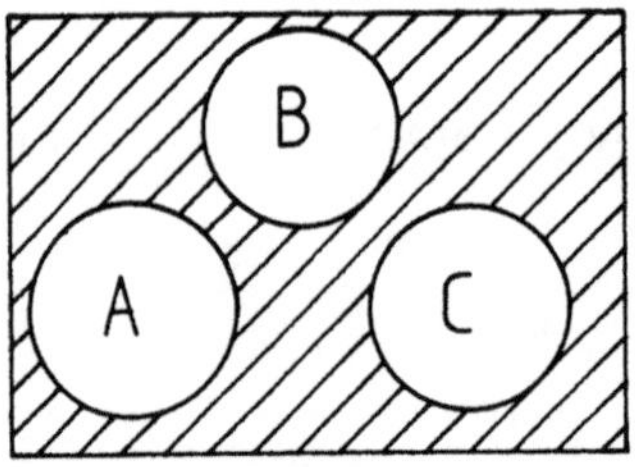

Figure 5.
Paths are mutually exclusive.

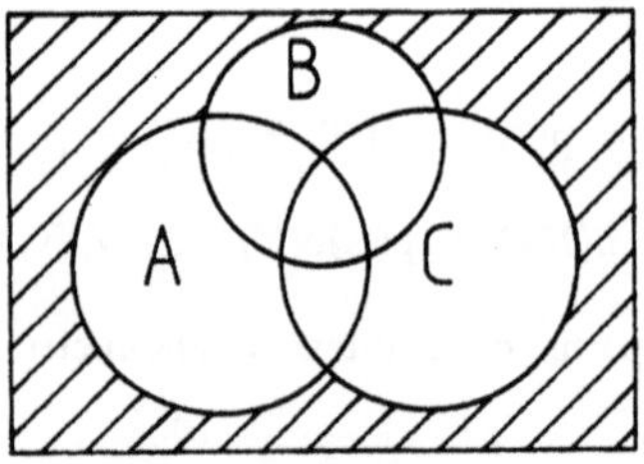

Figure 6.
Paths are independent or dependent to unknown degree

It is also clear that the value produced by equation 4., which considers the paths to be subassumptive, see figure 4, will be less than or equal to the values produced by equations 1–3.

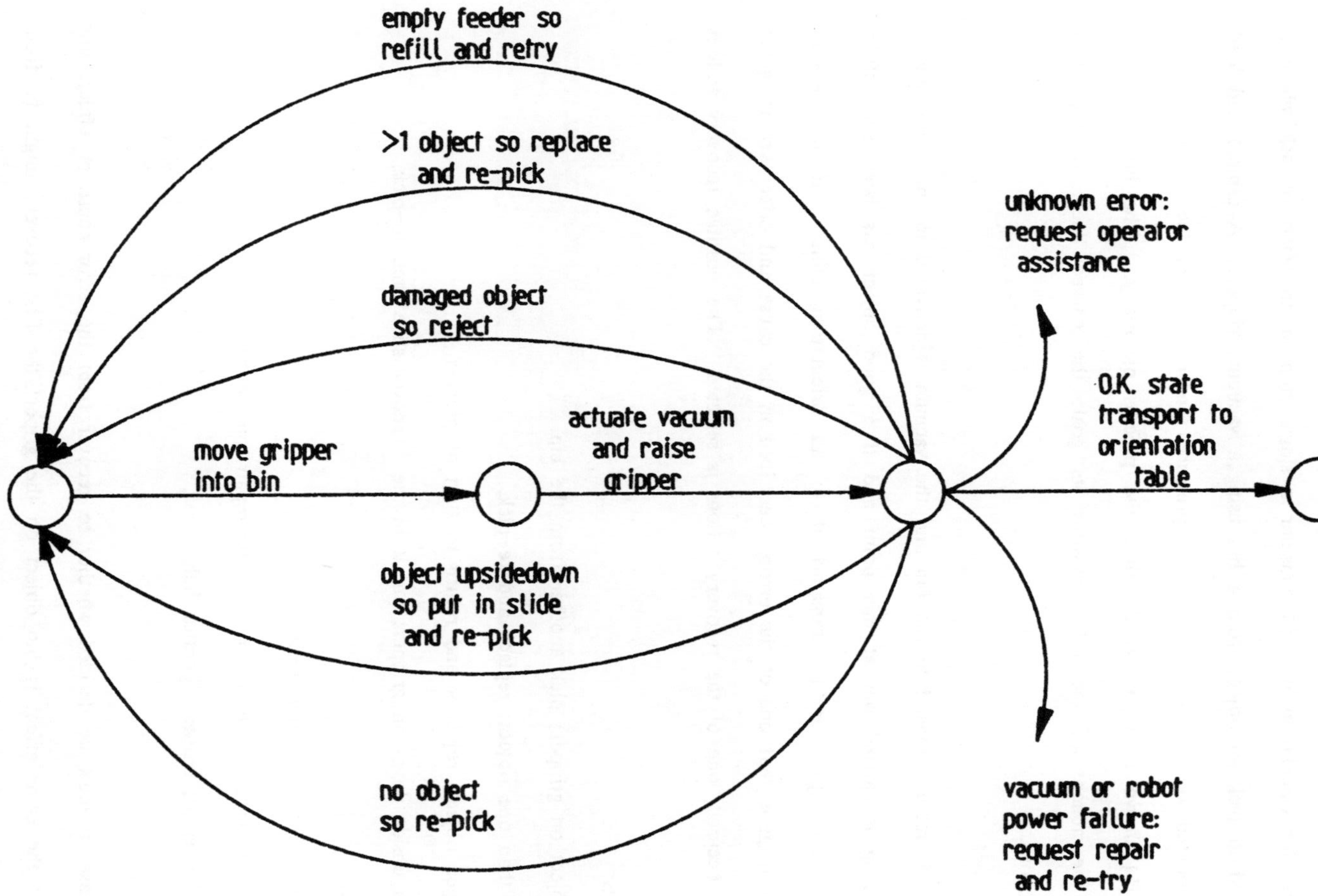

Figure 7. **Flowgraph of Picking Process.**

Example 1.

To illustrate this calculation we will consider a single step in an assembly task where a robot is used to pick an object from a bin using a vacuum gripper. A camera and load cell are mounted on the gripper, and a pressure sensor monitors the vacuum. The flowgraph [5] shown in figure 7. describes the process pictorially. The horizontal line represents the actions of the robot towards its goal; the curved lines are the error recovery actions.

The gripper is moved down into the bin and the vacuum is activated to 'pick' an object. Error checking is carried out at this point and if a good object has been successfully picked the robot goes on to transport it to an orientation table. If the sensory information suggests that one of the errors, described in the cause and effect vectors given below, has occured, one of the recovery actions is selected. The possible recovery actions are:

from effects:

no object on gripper; pick another from the bin.

more than one object; replace and re-pick.

damaged object; reject it and return to start of process.

object upside down on gripper; put it in the 'turnover slide' and re-pick.

from causes:

empty feeder; refill and re-pick.

vacuum or robot power failure; request repair and re-try.

unknown error; request operator help.

In some cases it must be decided whether to recover from the error cause or effect. For instance, if the error effect is 'no object on the griper' the first recovery might be from that effect. After repeated errors due to this effect the cause 'empty feeder' is recovered from. However, in other cases it is preferable to always recover from the error effect, as with the effect 'two objects on the gripper', where to recover from the underlying cause 'imperfect picking process' would require the redesign of the process. There are some cause-effect pairs which entail the same recovery action; recovering from the cause 'damaged object' and the effect 'damaged object on gripper' both necessitate discarding the object and picking another.

Note: The 'turnover slide' is a small chute which returns the objects to the bin, turning them over as they descend. It is possible to detect that an object is upside down using the camera because the two faces of the object have differing reflectances, resulting in an imperfect image of one side of the object, and thus different area and perimeter values for each side.

The forward calculation will be considered, from the error causes as far as the parameters affected. Hence, the probability vectors are

$$c = \begin{bmatrix} \text{empty feeder} \\ \text{vacuum failure} \\ \text{robot power failure} \\ \text{damaged object} \\ \text{picking process} \\ \text{O.K.} \\ \text{unknown error} \end{bmatrix}$$

$$e = \begin{bmatrix} \text{no object on gripper} \\ \text{>1 object on gripper} \\ \text{damaged oject on gripper} \\ \text{object upside down on gripper} \\ \text{O.K.} \end{bmatrix}$$

$$p = \begin{bmatrix} \text{vacuum incorrect} \\ \text{zero weight} \\ \text{non-zero, incorrect weight} \\ \text{incorrect object area} \\ \text{incorrect object perimeter} \end{bmatrix}$$

We may now define the transformation matrix C such that $e = C.c$. It is clear that there will be no object on the gripper if the feeder is empty or the vacuum fails, there may be a slight chance of an object remaining on the gripper, despite a robot power failure, due to the continued vacuum operation, and the imperfections of the picking process may mean that there is no object on the gripper. If there is no error (O.K. state) then there is no possibility of there being no object on the gripper, and there is a chance of there being no object on the griper due to some unknown error. Thus, the first row of the C matrix might be (1 1 0.8 0 0.1 0 0.3). For the second row we note that the only chance of there being more than one object on the gripper is through imperfections in the picking process. Thus we may select the second row to be of the form (0 0 0 0 0.2 0 0). Using similar arguments for the other rows we may determine the C matrix to be:

$$C = \begin{bmatrix} 1.0 & 1.0 & 0.8 & 0.0 & 0.1 & 0.0 & 0.3 \\ 0.0 & 0.0 & 0.0 & 0.0 & 0.2 & 0.0 & 0.0 \\ 0.0 & 0.0 & 0.0 & 1.0 & 0.0 & 0.0 & 0.0 \\ 0.0 & 0.0 & 0.0 & 0.0 & 0.5 & 0.0 & 0.0 \\ 0.0 & 0.0 & 0.0 & 0.0 & 0.0 & 1.0 & 0.2 \end{bmatrix}$$

In a similar manner we may define the E matrix, where $p = E*e$, as

$$E = \begin{bmatrix} 1.0 & 0.0 & 0.0 & 0.0 & 0.0 \\ 1.0 & 0.0 & 0.0 & 0.0 & 0.0 \\ 0.0 & 1.0 & 0.3 & 0.0 & 0.1 \\ 1.0 & 0.3 & 0.1 & 1.0 & 0.1 \\ 1.0 & 0.3 & 0.1 & 1.0 & 0.1 \end{bmatrix}$$

Noting that if there is no object on the gripper then the vacuum will be incorrect to obtain the first row. The same effect will also give us zero weight, as described by the second row, and so on for the other rows of *E*. We note that rows one and two and rows four and five are identical, suggesting that there is some redundancy in the sense parameter set used in this pedagogical example.

Using the * operation we can combine the *E* and *C* matrices to obtain a transformation between *c* and *p*.

$$E{*}C = \begin{bmatrix} 1.0 & 1.0 & 0.8 & 0.0 & 0.1 & 0.0 & 0.3 \\ 1.0 & 1.0 & 0.8 & 0.0 & 0.1 & 0.0 & 0.3 \\ 0.0 & 0.0 & 0.0 & 0.3 & 0.2 & 0.1 & 0.0 \\ 1.0 & 1.0 & 0.8 & 0.1 & 0.5 & 0.1 & 0.3 \\ 1.0 & 1.0 & 0.8 & 0.1 & 0.5 & 0.1 & 0.3 \end{bmatrix}$$

Thus if a vacuum failure were to occur, $c = (0\ 1\ 0\ 0\ 0\ 0\ 0)$, we see that it could be detected by the vacuum, zero weight and object area and perimeter parameters since

$$p = E{*}C \ . \begin{bmatrix} 0 \\ 1 \\ 0 \\ 0 \\ 0 \\ 0 \\ 0 \end{bmatrix}$$

$$p = (1\ 1\ 0\ 1\ 1)$$

Similarly if we were to suppose that all error causes were equally likely,

$c = (0.14 \quad 0.14 \quad 0.14 \quad 0.14 \quad 0.14 \quad 0.14 \quad 0.14)$, then

$$p = E*C \cdot \begin{bmatrix} 0.14 \\ 0.14 \\ 0.14 \\ 0.14 \\ 0.14 \\ 0.14 \\ 0.14 \end{bmatrix}$$

$$p = (0.45 \quad 0.45 \quad 0.09 \quad 0.53 \quad 0.53)$$

Note that while E and C are combined using the * operator, the result of this is combined with c using ordinary matrix multiplication because the elements of c are mutually exclusive.

The * operation is equivalent to the T–conorm $S_3(a,b)$ described by Bonissone in [3].

Upper bound ! operator

The operation ! between two matrices Y and Z is defined as

$$X = Y \; ! \; Z$$

$$x(i,j) = \min \left\{ \sum_k y(i,k).z(k,j), \; 1 \right\}$$

This produces an upper bound transformation matrix by summing the path lengths depicted in figure 3a. Thus T = p(A) + p(B) + p(C) gives an accurate result if the paths are mutually exclusive. If the paths are independent, dependent to an unknown degree (equations 2–3.), or subassumptive then ! produces a value larger than the correct value.

Example 2

Considering the same situation as example 1., we have the same c, e, p vectors and C and E matrices and can combine the two matrices using the ! operation to obtain

$$E!C = \begin{bmatrix} 1.00 & 1.00 & 0.80 & 0.00 & 0.10 & 0.00 & 0.30 \\ 1.00 & 1.00 & 0.80 & 0.00 & 0.10 & 0.00 & 0.30 \\ 0.00 & 0.00 & 0.00 & 0.30 & 0.20 & 0.10 & 0.02 \\ 1.00 & 1.00 & 0.80 & 0.10 & 0.66 & 0.10 & 0.32 \\ 1.00 & 1.00 & 0.80 & 0.10 & 0.66 & 0.10 & 0.32 \end{bmatrix}$$

It is clear that the only elements which have changed due to the change of combining function are those where there is more than one path between *c* and *p*. That is, there is more than one non zero element in a row of the *E* matrix which is multiplied by non zero elements in the columns of the *C* matrix. Some of the values in the matrix *E!C* are greater than those produced using the * function, showing how ! is an upper bound operator.

The parameter vector produced in response to a vacuum failure is the same as that in example 1. The parameter vector will only show a change if an incorrect object perimeter is represented in the cause vector. If all error causes are equiprobable the parameter vector obtained is

$$p = E!C \cdot \begin{bmatrix} 0.14 \\ 0.14 \\ 0.14 \\ 0.14 \\ 0.14 \\ 0.14 \\ 0.14 \end{bmatrix}$$

$$p = (0.45 \quad 0.45 \quad 0.09 \quad 0.56 \quad 0.56)$$

which shows an increase in the value of elements c_4 and c_5. Elements c_1, c_2 and c_3 were unchanged because there is only one path between the effect vector and the parameters involved, resulting in the upper and lower bounds being equivalent.

The ! operation is equivalent to the T–conorm $S_1(a,b)$ described by Bonissone in [3].

<u>**Best guess, #, operator**</u>

The operation # between two matrices $\boldsymbol{Y}$ and $\boldsymbol{Z}$ performs the following operation when the paths are considered (for a three path example).

$$T = p(A) + p(B) + p(C) - p(AB) - p(AC) - p(BC) + p(ABC)$$

That is, it represents the total area shown un-shaded in figure 3b.

This operator produces a transformation matrix which is between the upper and lower bounds produced by the operators * and !. No information is available about the inter-dependence of paths and so this operator assumes conditional independence. Cheeseman [4] states that, when no information is available, this is the only reasonable assumption which can be made, providing a 'neutral background' to the calculations. An accurate result is obtained if the paths are independent. The result is larger than the correct result if the paths are subassumptive, but less than the correct result if they are mutually exclusive.

The best guess # operator is equivalent to T-conorm $S_2(a,b)$ described by Bonissone in [3].

<u>Example 3</u>

Considering the same arrangement used in example 1., we have the same vectors and matrices. Combining the $\boldsymbol{E}$ and $\boldsymbol{C}$ matrices using # we obtain

$$\boldsymbol{E\#C} = \begin{bmatrix} 1.00 & 1.00 & 0.80 & 0.00 & 0.10 & 0.00 & 0.30 \\ 1.00 & 1.00 & 0.80 & 0.00 & 0.10 & 0.00 & 0.30 \\ 0.00 & 0.00 & 0.00 & 0.30 & 0.20 & 0.10 & 0.02 \\ 1.00 & 1.00 & 0.80 & 0.10 & 0.58 & 0.10 & 0.31 \\ 1.00 & 1.00 & 0.80 & 0.10 & 0.58 & 0.10 & 0.31 \end{bmatrix}$$

As explained in example 2. only elements in the fifth column of this matrix have changed with the use of this operator, and their values are seen to be between the values produced by the upper and lower bound operators. If all the error causes are equiprobable then the parameter vector obtained is p = (0.45 0.45 0.09 0.54 0.54). The probability of the viewed object's perimeter being incorrect due to the imperfections of the picking process, according to the use of the three different operators, is

$$0.5 < p(\text{perimeter incorrect}) < 0.66$$

and the best guess is 0.58.

We see that

$$_{*}T < _{\#}T < _{!}T$$

If the paths are dependent to an unknown degree then $_{\#}T$ could be greater than or less than $_{true}T$

Note: $_{*}T$, $_{\#}T$, $_{!}T$ represent the E and C matrices combined using the subscripted operator.

$_{true}T$ represents the ideal value of T which could be calculated if all dependencies were known.

INVERSE CALCULATION

In the inverse calculation information from the sensors is used to identify the most probable error cause. A simple procedure which serves to illustrate the process is described below, and uses the † operator, which assumes that all the matrix entries are 0 or 1.

Inverse, †, operator

The sensor vector s is determined during run time, with s_1 = 1 indicating an active sensor and 0 indicating an inactive sensor. In the case

$$e^{T} = p^{T} \dagger E$$

the operation of † is defined as follows

$$\begin{aligned} e_i &= 1 \quad \text{if} \quad p_j = e_{ji} \qquad \forall j \\ &= 0 \quad \text{otherwise} \end{aligned}$$

Any vector elements which are unknown (due to incomplete sensing) are given the value X. X is treated as a match for 1 or 0. This means that the † operation produces an upper bound result, since it searches for all possible error causes with the limited information available. (Note: A sensor vector with all X's would produce an error vector full of 1's).

If the † operator is used with X matching neither 1 or 0 it produces a form of lower bound. The result is that full sensing is required before the operator will give any results.

<u>Example 4</u>

As an example the error cause will be deduced for the parameter vector. The c and p vectors are of the same form as in example 1. The E and C matrices have their elements rounded to unity or zero and hence become

$$C = \begin{bmatrix} 1 & 1 & 1 & 0 & 0 & 0 & 0 \\ 0 & 0 & 0 & 0 & 0 & 0 & 0 \\ 0 & 0 & 0 & 1 & 0 & 0 & 0 \\ 0 & 0 & 0 & 0 & 1 & 0 & 0 \\ 0 & 0 & 0 & 0 & 0 & 1 & 0 \end{bmatrix}$$

$$E = \begin{bmatrix} 1 & 0 & 0 & 0 & 0 \\ 1 & 0 & 0 & 0 & 0 \\ 0 & 1 & 0 & 0 & 0 \\ 1 & 0 & 0 & 1 & 0 \\ 1 & 0 & 0 & 1 & 0 \end{bmatrix}$$

Note that elements $c(2,5)$, $c(1,7)$ $c(5,7)$ have been rounded down to 0,making row 2 and column 7 redundant in the C matrix. Similarly column three of E is redundant, illustrating the limitations whlich the rounding process imposes. The p vector is given as p = (1 X X X X), meaning that the vacuum was found to be incorrect and the other parameters had not been sensed. Thus

$$e^{T} = p^{T} \dagger E$$

$$e^T = [1\ \ X\ \ X\ \ X\ \ X] \quad \dagger \quad \begin{bmatrix} 1 & 0 & 0 & 0 & 0 \\ 1 & 0 & 0 & 0 & 0 \\ 0 & 1 & 0 & 0 & 0 \\ 1 & 0 & 0 & 1 & 0 \\ 1 & 0 & 0 & 1 & 0 \end{bmatrix}$$

$$e^T = [1\ \ 0\ \ 0\ \ 0\ \ 0]$$

meaning that there was no object on the gripper. It is clear that each column of E relates to a different error effect, and the p vector is compared with each of the columns. Now

$$c^T = e \dagger C$$

$$c^T = [1\ \ 0\ \ 0\ \ 0\ \ 0] \quad \dagger \quad \begin{bmatrix} 1 & 1 & 1 & 0 & 0 & 0 & 0 \\ 0 & 0 & 0 & 0 & 0 & 0 & 0 \\ 0 & 0 & 0 & 1 & 0 & 0 & 0 \\ 0 & 0 & 0 & 0 & 1 & 0 & 0 \\ 0 & 0 & 0 & 0 & 0 & 1 & 0 \end{bmatrix}$$

$$c^T = [1\ \ 1\ \ 1\ \ 0\ \ 0\ \ 0\ \ 0]$$

which means that the error cause could be any one of:

empty feeder

vacuum supply failure

robot power failure

'X' is treated as a match for '1' or '0' , meaning that † gives an upper bound result. Using 'X' to match only 1's or only 0's gives only a subset of the possible error cause vector.

This operator is seen as being useful for quickly reducing the set of possible error causes or effects, and the invoking of sensing to further reduce the set, prior to a more detailed calculation.

<u>Bayesian Inverse Calculation</u>

One method of performing the inverse calculation and thus determining the error cause from the sensory information uses Bayes theorem in the form

$$p(c_i | s_j) = \frac{p(s_j | c_i) . p(c_i)}{\sum_k p(s_j | c_k) . p(c_k)} \tag{5}$$

This requires the forward calculation to have been done to obtain the conditional probabilities $p(s_j | c_i)$. The a priori probabilities of the error causes are also needed. The forward calculation can be done using any of the operators *, ! or # and will produce a bound or best guess result accordingly. When a sensor becomes active equation 5. is used to calculate the a posteriori probabilities of the various error causes.

The summation in the denominator of equation 5. requires that the error causes are mutually exclusive and exhaustive. For this reason, the analysis is restricted to the occurrence of a single error at a time. If two sensors should be active simultaneously, then the a posteriori c vector is first calculated using s_1, the resultant a posteriori c vector is then used as the a priori c vector in the calculation of the a posteriori c vector using s_2. This results in a c vector containing values which represent the possible causes for the error sensed by s_1 <u>and</u> s_2. Alternatively, an expanded form of Bayes theorem may be used which calculates the posterior error cause vector, based on both pieces of sensing information, simultaneously.

<u>Example 5</u>

If the $\boldsymbol{E}$ and $\boldsymbol{C}$ matrices used in example 1. are combined using, say, the # operator, we obtain

$$E\#C = \begin{bmatrix} 1.00 & 1.00 & 0.80 & 0.00 & 0.10 & 0.00 & 0.30 \\ 1.00 & 1.00 & 0.80 & 0.00 & 0.10 & 0.00 & 0.30 \\ 0.00 & 0.00 & 0.00 & 0.30 & 0.20 & 0.10 & 0.02 \\ 1.00 & 1.00 & 0.80 & 0.10 & 0.58 & 0.10 & 0.31 \\ 1.00 & 1.00 & 0.80 & 0.10 & 0.58 & 0.10 & 0.31 \end{bmatrix}$$

Using equation 5, with p($\boldsymbol{p}$|$\boldsymbol{c}$) replacing p($\boldsymbol{s}$|$\boldsymbol{c}$), and the a priori $\boldsymbol{c}$ vector, $\boldsymbol{c}$ = (0.2 0.07 0.07 0.15 0.24 0.2 0.07), it is possible to calculate the matrix B which represents p($\boldsymbol{c}$|$\boldsymbol{p}$), the 'backward transformation' between the sense parameters and the error causes, such that $\boldsymbol{c} = B.\boldsymbol{p}$. We obtain

$$B = \begin{bmatrix} 0.54 & 0.54 & 0.00 & 0.38 & 0.38 \\ 0.19 & 0.19 & 0.00 & 0.13 & 0.13 \\ 0.15 & 0.15 & 0.00 & 0.11 & 0.11 \\ 0.00 & 0.00 & 0.39 & 0.03 & 0.03 \\ 0.06 & 0.06 & 0.42 & 0.27 & 0.27 \\ 0.00 & 0.00 & 0.17 & 0.04 & 0.04 \\ 0.06 & 0.06 & 0.01 & 0.04 & 0.04 \end{bmatrix} = p(\boldsymbol{p}|\boldsymbol{c})$$

If an error in the vacuum pressure is detected, giving the sense parameter vector $\boldsymbol{p}$ = (1 0 0 0 0), the a posteriori error cause vector, $\boldsymbol{c}$, is calculated in the following manner.

$$\boldsymbol{c} = B.\boldsymbol{p}$$

$$\boldsymbol{c} = \begin{bmatrix} 0.54 & 0.54 & 0.00 & 0.38 & 0.38 \\ 0.19 & 0.19 & 0.00 & 0.13 & 0.13 \\ 0.15 & 0.15 & 0.00 & 0.11 & 0.11 \\ 0.00 & 0.00 & 0.39 & 0.03 & 0.03 \\ 0.06 & 0.06 & 0.42 & 0.27 & 0.27 \\ 0.00 & 0.00 & 0.17 & 0.04 & 0.04 \\ 0.06 & 0.06 & 0.01 & 0.04 & 0.04 \end{bmatrix} . \begin{bmatrix} 1 \\ 0 \\ 0 \\ 0 \\ 0 \\ 0 \\ 0 \end{bmatrix}$$

The a posteriori c vector is thus, c = (0.54 0.19 0.15 0.00 0.06 0.00 0.06), that is, the probability of the error cause being:

empty feeder	=	0.54
vacuum failure	=	0.19
robot power failure	=	0.15
damaged object	=	0.00
imperfect picking process	=	0.06
O.K. - no error, carry on	=	0.00
unknown error cause	=	0.06

from which we see that the most likely error cause is an empty feeder.

Because there is only a single non-zero entry in the p vector, all the operators (*, !, #, and ordinary matrix multiplication) used to combine B and p produce the same c vector.

If the ! or * operators had been used to combine the E and C matrices, instead of #, then the a posteriori c vector obtained would have represented an upper or lower bound, respectively.

The B matrix was used to represent the transformation between the sense parameters and the error causes. More usefully, the matrices B_c and B_e are formed to transform between sensory information and the error causes, $c = B_c.s$, or the error effects, $e = B_e.s$.

CONCLUSION

The use of the probabilistic calculation system described here is proposed, offline, for sensor selection; online, for error diagnosis, within an error recovery tree, to direct further sensing or an error recovery action.

The mathematical framework described, together with appropriate knowledge bases, which are updated at run-time, aim to provide a workcell with an autonomous error control capability, which is necessary if sensory assembly tasks are to be economically competetive for small batch production quantities.

REFERENCES

[1] Taylor, G.E., Taylor, P.M., "Dynamic error probability vectors: a framework for sensory decision making." *IEEE Proceedings of Int. Conf. on Robotics and Automation*, Philadelphia. 1985. pp 1096 - 1100.

[2] Taylor, P.M., "Multisensory assembly and error recovery." (Proceedings of a NATO Advanced Reserach Workshop, Castleveccio di Pascoli, October 1987), in *Kinematic and dynamic issues in sensor based control*. Edited by G.E. Taylor. Springer Verlag. To be published 1990.

[3] Bonnissone, P.P., "Summarizing and propagating uncertain information with triangular -norms." *Int. J. of Approx Reasoning*. 1987: 1: pp 71-101. Elsevier.

[4] Cheeseman, P., "In defense of probability." *Procs. 9th Int. Joint Conf. on A.I.* 1985, Los Altos Calif. Kauffman. pp 1002 - 1009.

[5] Monkman, G.J., Taylor, G.E., Taylor, P.M., "Flowgraph techniques in workcell assessment and design". *IEEE International Symposium on Intelligent Control*, Albany, New York. September 1989. pp 28 - 32.

ADAPTIVE CONTROL

P.E. Wellstead
Control Systems Centre
UMIST
Manchester M60 1QD
England

ABSTRACT

The paper discusses some issues as they concern adaptive control of manipulators. In particular attention is focussed upon self-tuning digital control techniques and variations upon this scheme. A review of adaptive robot control is then presented and various research directions noted. Attention is then specifically addressed to the problem of self-tuning servo-control of manipulator joints. Finally the results of some practical applications are noted.

1. INTRODUCTION

The topic of adaptive control has passed through numerous cycles of alternating activity and passivity as a research topic. At the moment we seem to be at the peak of a period of intense activity both in terms of research and application of adaptive techniques. In particular, adaptive control is an extremely active research and application area with numerous industrial products available. On a broader plane, the principle of adaption has found application in all areas of algorithmic signal processing. To elaborate further, the research aspects of adaptive systems flow from several differing but

NATO ASI Series, Vol. F 64
Sensory Robotics for the Handling of Limp Materials
Edited by P. M. Taylor

complementary signal processing technologies. These include -

i) Control engineering.

ii) Prediction and forecasting (Wittenmark, 1974; Wellstead et al, 1987).

iii) Signal filtering (Widrow & Stearns, 1987).

In the context of this paper we will review results associated with adaptive control engineering. However, even this restriction of scope leaves an enormous potential range of activity. In particular, within the control engineering discipline itself numerous theoretical approaches have been adopted and differing application areas explored. Clearly, the interests of this workshop lie in robotic manipulation, and to this end the paper will concentrate on applications appropriate to adaptive robot control. The situation with theoretical approaches to adaptive control is not so clear-cut, with numerous rival procedures being suggested by various research groups. To help in this respect, an attempt will be made in this article to offer some guidance concerning the relative strengths of the various technical approaches. A further aspect of adaptive control theory concerns the convergence of adaptive algorithms. This topic will not be treated, however interested readers are refered to the texts by, Ljung & Soderstrom [1983], Goodwin & Sin [1984] and Anderson et al, [1986]

The reader should note that adaptive control is only one of the feasible solutions to the regulation of robot manipulator dynamics. The key problem with robot dynamics concerns their non-linear nature. One way of handling this is by a non-linear feedback which renders the overall system linear. A good example of this is in Freund [1982]. A similar argument suggests that gain scheduled control methods can be applied to robot control with response characteristics which are comparable with adaptive algorithms. Likewise, robust control methods are often proposed in robots. In reply to these points and without recourse to detailed discussion, it is possible to state that adaptive controls have been found to have excellent practical properties by many robot researchers. An indication of these properties will be given in section 3. Moreover, the central reason for presenting adaptive control concepts in the current context is to motivate their use in the handling of limp materials.

2. APPROACHES TO ADAPTIVE CONTROL

Adaptive control has, as noted earlier, been through various cycles of activity. The current cycle of activity commenced around the early 1970's with the introduction of the 'self-tuning control' concept, [Astrom & Wittenmark, 1973]. The key feature of a self-tuning controller is the combination of a recursive parameter estimator with a controller synthesis rule in a real-time control algorithm (Fig.1a). The self-tuning control cycle consists of a set of algorithms which:-

a) Update estimates of a recursive model of the system to be controlled.

b) Use these new estimates to re-synthesize the parameters of a control law.

c) Compute and apply a control input on the basis of the updated control law.

This sequence is repeated at each sample interval.

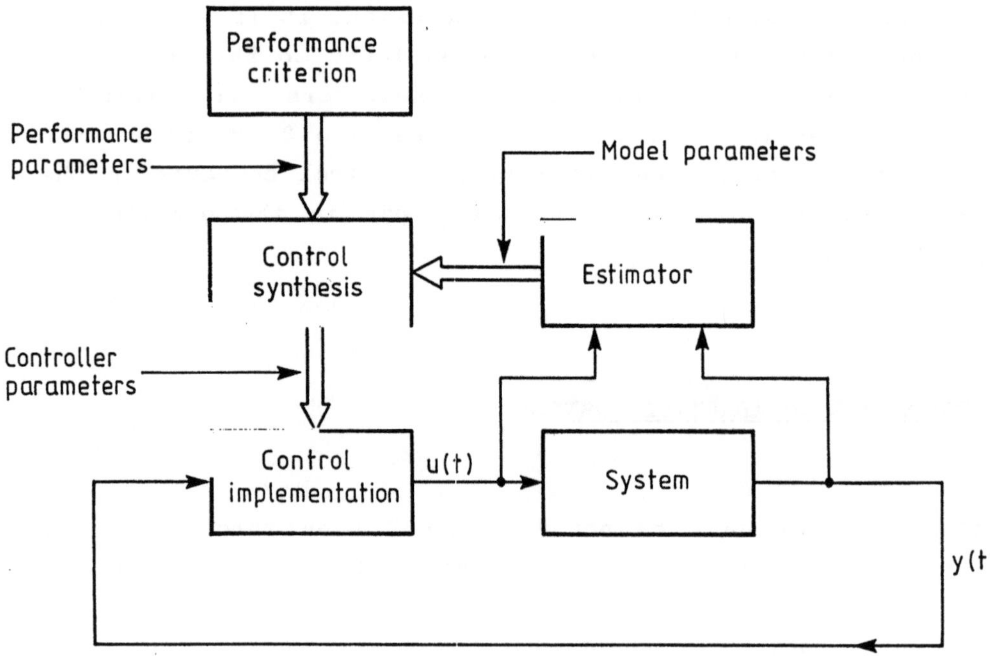

Figure 1a. The General Schema of a Self-Tuning Regulator (STR) System.

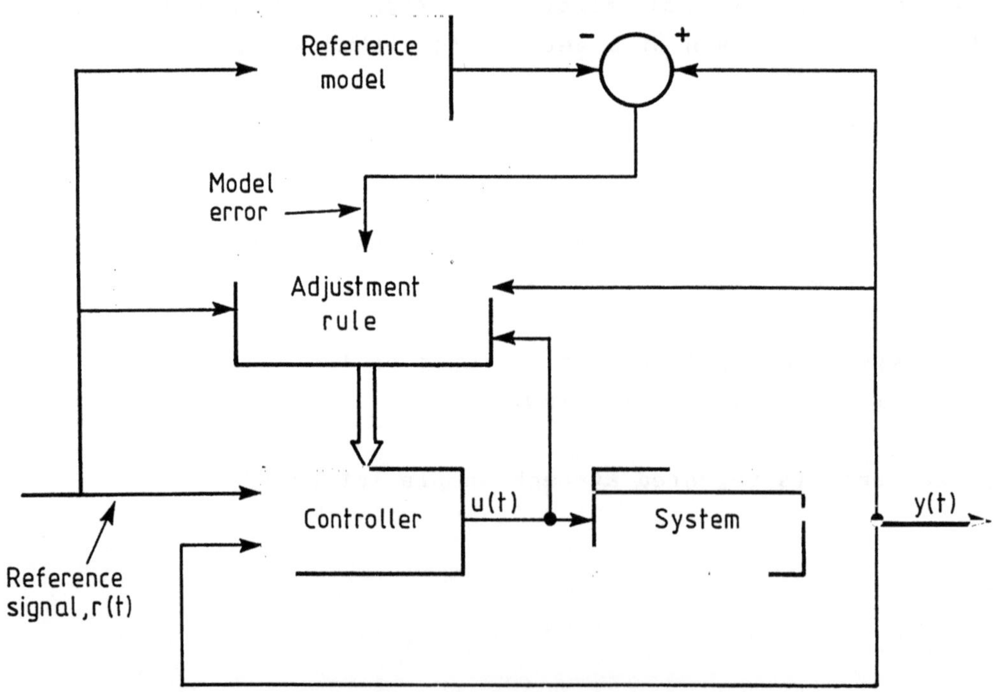

Figure 1b. The General Schema of a Model Reference Adaptive System (MRAS).

The intuitive appeal of this form of adaptive controller has been such that alternate versions, such as Model Reference Adaptive Control (MRAS) (Fig.1b) have diminished in popularity in many areas of application. A particular instance is the process control area where the self-tuning control approach is overwhelmingly used [see, eg. Seborg et al, 1986]. Within the robot adaptive control sub-culture the situation is more equally divided between MRAS and STR algorithms. Not withstanding this the overview presented below will focus on self-tuning algorithms. In any event it is possible to show formal equivalences between self-tuning and MRAC methods. For a current survey of Model Reference Adaptive Control see [Astrom, 1987] and the text book [Landau, 1979]. By the same token, section 3 will mention some MRAS applications in robot control.

2.1. Self-Tuning Estimators

An essential component of a self-adaptive control algorithm is a recursive estimator. The role of this object is to obtain parameter estimates for a model of the underlying system. Numerous estimation/identification algorithms are used in practice. The precise form of identifier will depend upon the goal of the adaptive algorithm and the type of control law which is to be embedded in the synthesizer. For example, in adaptive spectral equalization a frequency response estimator is used together with a frequency domain control synthesis rule [eg. Wellstead & Zarrop, 1983]. Correspondingly, in three term control auto-tuners [eg. Kraus and Myron, 1984] a simple time domain estimator is used based upon transient response measurement.

However, despite these variations the mainstream of adaptive control is based around recursive parameter estimation of a linear model of the form

$$A\,y(t) = B\,u(t) + v(t) \tag{1}$$

where, y(t), u(t) are sequences of system output and input variables respectively measured at controller sampling intervals. The variable v(t) is a disturbance process which may or may not be measurable.

The objects A,B are polynomials of the form

$$\left.\begin{aligned} A &= 1 + a_1x + a_2x + \ldots + a_{n_a}x^{n_a} \\ B &= b_1x + \ldots + b_{n_b}x^{n_b} \end{aligned}\right\} \tag{2}$$

in which x is either the backward shift operator z^{-1} or the δ operator defined by

$$\delta = h^{-1}(1 - z^{-1}) \tag{3}$$

where h^{-1} is the sample interval.

The z^{-1} operator is widely used in process and industrial control applications, whereas δ operator methods are more applicable to servo control in which fast sampling is required.

There are many different methods which could be used for estimating the parameters of A,B. For example, stochastic approximation, least squares and its variants [Lawson & Hanson, 1974], instrumental variables and maximum likelihood. A review of these procedures is given in, for example, [Norton, 1986]. There is no absolutely best recursive estimator for self-adaptive control. However, the basic and most widely used algorithm is recursive least squares.

In recursive least squares estimation the criterion

$$V = \sum_{i=1}^{t} (\hat{A}y(t) - \hat{B}u(t))^2 \qquad (4)$$

is minimised. There are various recursive formulations for performing this minimisation. The traditional approach is via the Matrix Inversion Lemma, however square root algorithms are used widely where numerical errors are a concern [Bierman, 1977]. Fast recursions are useful where either A or B is of high degree (eg: n_a or n_b > 10), but parallel processing recursions based, for example upon the Given's type transformation have all but replaced the need for fast algorithms. [Gentleman, W., 1974].

2.2. Controller Synthesis Methods

The main area in which adaptive controllers differ from one another is in the selection of controller synthesis methods. This selection is strongly influenced by the objective of the controller but it must also be said that personal prejudice on the part of the individual researcher is an influential factor. In the following paragraphs, an attempt will be made to outline the principal controller synthesis rules, and to place them into an appropriate engineering context.

Minimum Variance Control [Astrom & Wittenmark, 1973]

This is a stochastic control approach in which the disturbance v(t) is assumed to be an unknown random process. The controller then attempts to minimize the cost

$$J = E[y^2(t) + \lambda\ u^2(t)] \qquad (5)$$

where λ is a weighting coefficient on the control action, and E["] is the expectation operator.

The key use of such an approach is in stochastic regulation where setpoint changes are infrequent or well defined, and where the optimisation of a cost object (equation (5)) is a sensible control objective.

Linear Quadratic Control [Kárný et al, 1985]

Minimum variance control cancels the system zeros and will therefore fail when the system is non-miminum phase. An alternative stochastic controller which avoids this is Linear Quadratic (LQ) control which minimizes

$$J_2 = \lim_{N \to \infty} E \left\{ \frac{1}{N} \sum_{t=1}^{N} (y^2(t) + \lambda\ u^2(t)) \right\} \qquad (6)$$

This algorithm requires the solution to a steady state Riccati equation, this is computationally quite demanding compared to the minimum variance technique.

Predictive Control **[Clarke et al, 1987]**

An alternative to LQ based control which can still avoid the problems of minimum variance control is the class of predictive control algorithms. This approach minimizes a cost function of the form.

$$J_3 = E\left\{ \sum_{j=0}^{N} (y^2(t+j) + \sum_{j=0}^{N} \lambda\ u^2(t+j) \right\} \qquad (7)$$

This form of cost is descended from the minimum variance type cost (equation 5) and attempts to moderate current control actions by weighting them in some way which is related to future behaviour.

Pole-Placement Control **[Wellstead et al, 1979]**

The control procedures outlined above have all concerned the optimisation of some stochastic cost function. As such the acheived closed loop dynamical performance is only indirectly linked to the design criterion. An alternative approach is to select a design/synthesis criterion which directly determines the closed loop dynamical behaviour. An approach which meets this need is pole-assignment or pole-placement self-tuning. The idea here is to select a control law:

$$F\ u(t) = H\ r(t) - G\ y(t) \qquad (8)$$

such that the F,G controller polynomials are determined by equating the actual closed loop characteristic equation with a desired characteristic polynomial T. Thus

$$F\ A + B\ G = T \tag{9}$$

where T is user selected to define the desired closed loop pole set. The polynomial H is a precompensator selected to define the desired response to setpoint changes.

Pole-placement is thus a classically derived algorithm, moreover it does not suffer from the shortcomings of minimum variance control. The solution to the identity is conducted on-line by either direct matrix inversion or via polynomial reduction techniques (Kučera, 1979).

The pole-placement approach can have stochastic disturbance reduction built into it if required (Zarrop & Fischer, 1985]. Moreover, optimal algorithms can be given pole-placement properties. [eg. Allidina & Hughes 1980] and [Lelic and Zarrop, 1986].

3. APPLICATION OF ADAPTIVE CONTROL IN ROBOTICS.

3.1. General Application Review.

Before discussing specific issues, we briefly note some significant and useful contributions to adaptive control of robots. The usual approach to adaptive robot control is to model the dynamical behaviour by an approximate linear representation. In this approach the non-linear behaviour of the robot is accounted for in quasi-equivalent time-variations in the model parameters. It is subsequently the task of the recursive estimator to track these time varying parameters and compensate for them in the control action. An early use of this approach is given in Leininger & Wang [1982]. In the same spirit, the work of Dubowsky [1979] is interesting for comparative purposes since it is an application of MRAC to robot control. Likewise, Koivo & Guo [1981] is typical of the optimal control based self-tuning controllers as applied in robotics. Elliot, et al, [1984] use pole-assignment self-tuning concepts for robot control, but incorporate explicitly the non-linear nature of manipulator joint dynamics.

An interesting recent application of MRAS to manipulator control is given in [Das & Loh, 1987], these authors note that certain model reference formulations can experience convergence problems when used in sampled data form. The Das and Loh solution is to use a hybrid continous time/discrete control law. Similar results would be found if the delta operator model had been used as defined in equation 3.

Further MRAS applications in robot control are given in [Murphy & Saridis, 1987; Anwar et al, 1986; Horowitz and Tomizuka (1982)]. All of these approaches are based upon a discrete time linear model of the robot joint dynamics. Thus, as in Leininger & Wang [1982], the non-linearities are tracked as time varying parameters.

A different approach to adaptive robot control is to use the known structure of the problem or to emulate (in adaptive form) an existing robot control procedure. For example, Bastin, Guillaume & Campion [1987] use a self-tuning control approach with an adaptive Computed Torque Controller for robotics. Other workers place different control forms into an adaptive format. In this context the sliding-mode adaptive methods of Slotine & Coetsee [1986] are significant. (see also Slotine & Weiping [1987]). Other concerns of robotic researchers have focussed largely on the question of unmodelled dynamics. In this context, adaptive algorithms (MRAS or STR) are known to have problems when the system model used in adaption does not cover the system dynamics. In robotics this can be a problem when the flexural modes of the robot are significant. (see for example, Yuh, 1987).

3.2. Self-Tuning Robotic Servo Control.

In this section we consider the servo control of robot manipulator joints using self-tuning algorithms. The aim here will be to illustrate typical results emanating from real applications and as a precursor to more general discussions of robot adaption. Because the robot problem is essentially one of servo-tracking, we will concentrate upon the application of pole-placement type methods. As a preliminary to this however it is useful to review the context and objectives of robot joint control in general.

The aim of a robot joint angle controller in general is to ensure that the angle y(t) follows a set-point r(t) determined from some trajectory specification. In this connection, the requirement is for good rise time and minimal overshoot. A further requirement is that the self-adaptive algorithm should have good steady state tracking performance (for the purposes of repeatability). In addition, there should be disturbance rejection facilities for handling load and geometry changes. Indeed a major reason for using self-tuning control in robotics is the ability of such algorithms to re-tune to changes in the underlying system.

The requirement for steady state tracking in practice means that the adaptive algorithm should incorporate integral action. Likewise, the requirement for direct control over transient response behaviour again suggests that a pole-assignment algorithm is most appropriate, since closed loop transient response shaping is possible by selection of T and H in equations (8,9).

The use of pole-assignment self-tuning for robotics has been extensively studied by Wahab [Wahab, 1985; Wahab & Wellstead, 1986; Wahab & Wellstead, 1989]. The application of generalized predictive self-tuning with pole-placement incorporated has been studied by Lelic [Lelic & Wellstead, 1987]. The discussion which follows relates to practical conclusions which were drawn from the studies of Wahab & Lelic upon the three principal axes of the Control Systems Centre (CSC) research robot. The CSC robot (Fig.2) is a revolute joint manipulator powered by d.c. motors through harmonic gearboxes. A full description of this hardware and the associated control software is given in [Wahab, 1985].

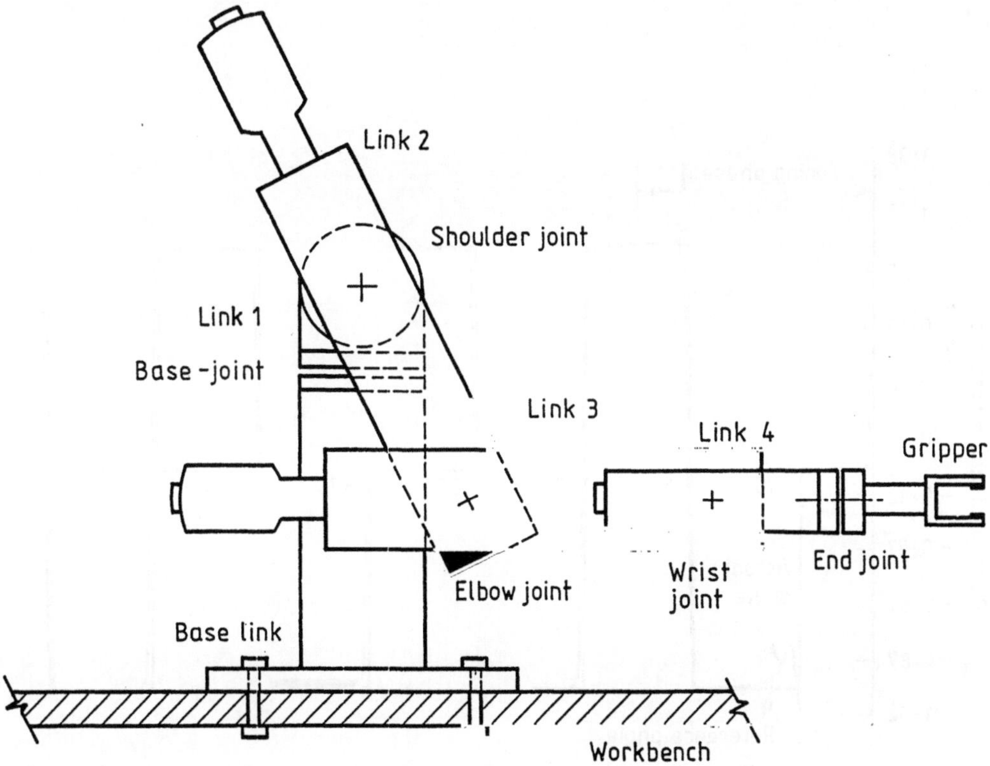

Figure 2. The Control Systems Centre (CSC) Research Robot.

3.2.1. Extended Pole-Assignment Control [Wahab, 1985]

The use of pole-assignment on each individual joint of the 3 principal axes proved to be highly successful. In general the adaptive controller tuned rapidly to the robot dynamics, in this context it typically took two setpoint changes for the controller to converge to its final configuration. Figure 3 shows a typical tuning transient for the base joint of the robot. The major contribution of the adaptive controller was found to be its initial auto-tuning ability; such that subsequent changes in load and joint position did not require significant adaption of the controller parameters. Likewise, a part of the adaptive control exercise conducted by Wahab concerned the use of multivariable self-adaptive control to reduce interactions.

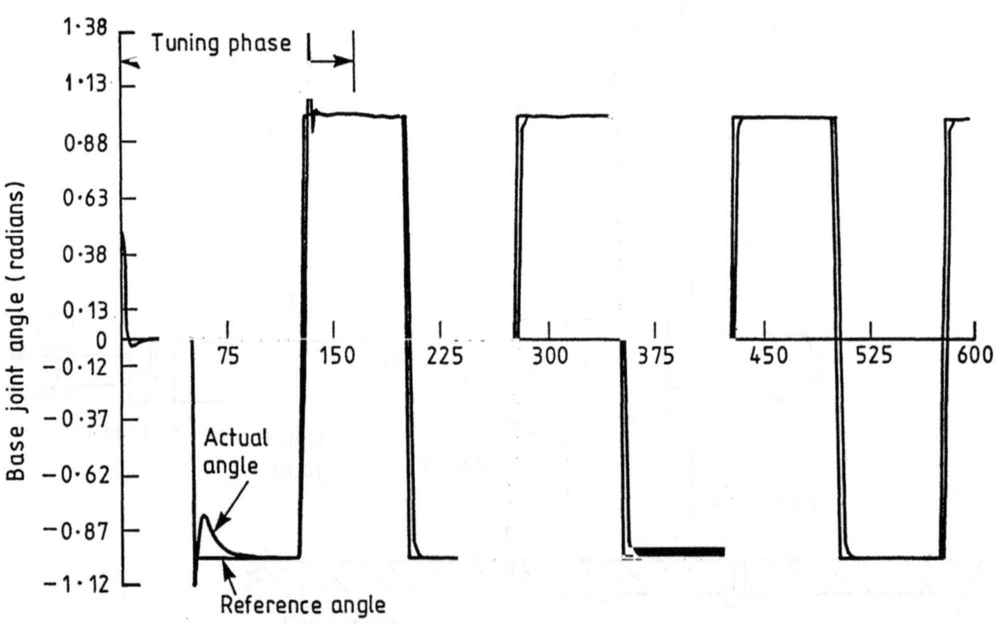

Figure 3. The self-tuning control of the base joint of the CSC research robot under pole-assignment control.

At a technical level the use of multivariable pole assignment [Prager & Wellstead, 1981] was proven to be possible. From a practical viewpoint however it is doubtful whether it is justified for a properly designed robot. The practical disadvantages concern the greatly increased computation time of the multivariable adaptive algorithms. This drawback should be weighed against the minimal improvement in performance associated with the reduced joint interactions which is achieved with multivariable control applied to a dynamically balanced manipulator.

3.2.2. Generalized Pole-Placement Control [Lelic & Wellstead, 1987]

The generalized pole-placement (GPP) algorithm is a member of the predictive self-adaptive controllers mentioned in section 3. Lelic's aim was to explore the benefits of such adaptive methods in robots. A particular feature of the GPP method is that it allows future knowledge of the setpoint to be programmed into current control actions. In a manipulator with a pre-programmed trajectory cycle this is clearly an algorithmic feature which can be exploited. For example, Figure 4 compares the response of a single joint under GPP self-adaption with and without use of predictive setpoint. Notice that the output response is similar in either case, but that the control action is reduced when setpoint programming is used. This smoother control action has implications for actuator wear and the power rating of actuators.

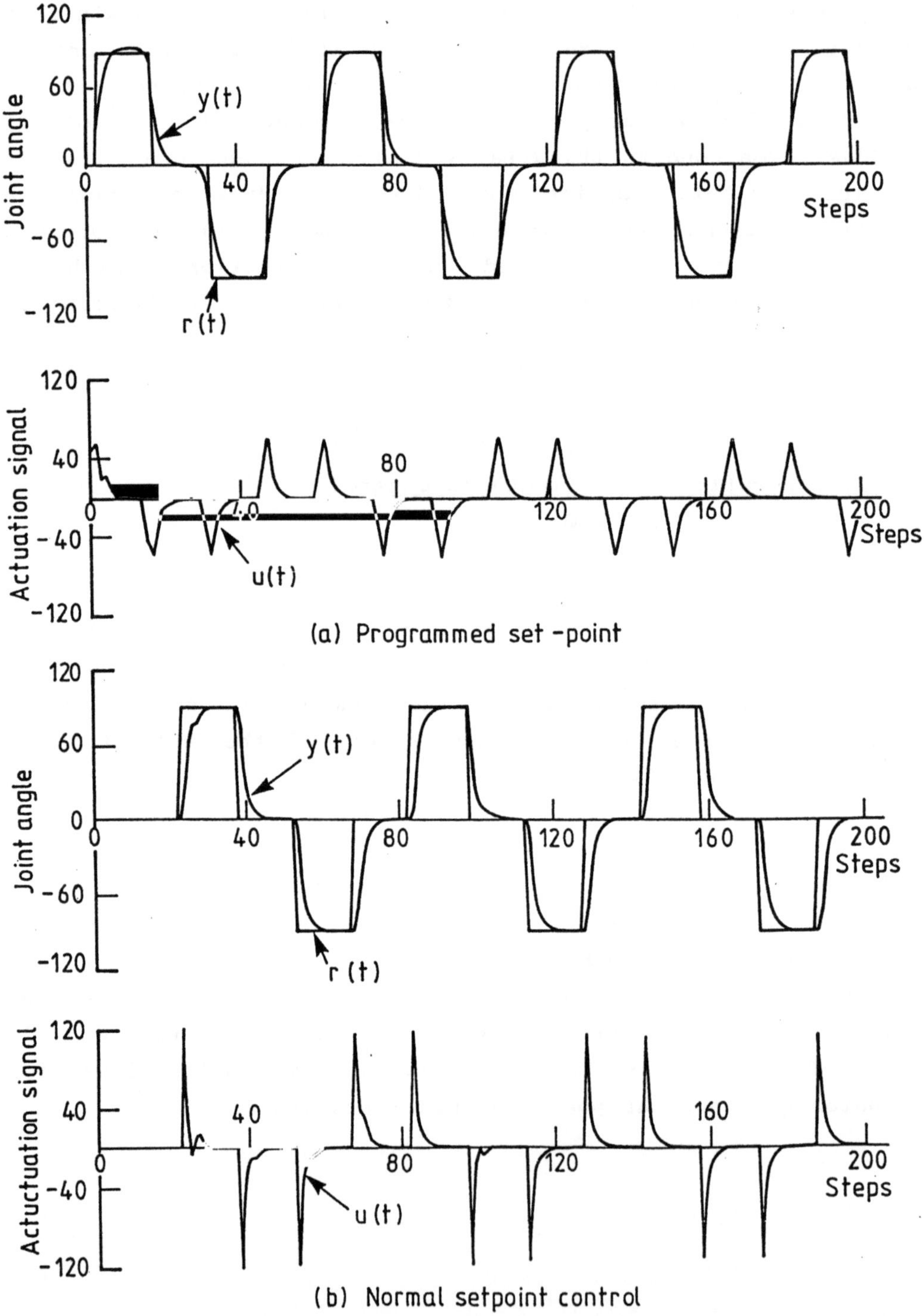

Figure 4. The self-tuning control of the shoulder joint of the CSC research robot under GPP control, comparing the uses of programmed setpoint control with normal servo-control.

4. FURTHER DEVELOPMENTS

This overview has been written from a control theoretic perspective. In this respect, I have tried to provide background on the range of adaptive algorithms which exist, together with some practical insights into their application for joint control. It should be reaffirmed that the practical studies discussed in section 3 referred to a well designed, rigid body robot. Moreover, the robot is intended to work in a fixed environment without significant interaction with the work-piece. Open issues which flow from current experience concern what happens when these conditions do not apply. For example, the work on flexible robots cited previously (see also Sidman & Franklin, 1987) comes towards meeting the unmodelled dynamics problem in robotics, although theoretical difficulties remain with adaptive control of incompletely modelled systems.

With regard to the subject of this meeting it seems that a more important question would concern the use of adaptive control in interactions with the work piece, or indeed other robotic devices [eg. Slotine & Li, 1987]. Unfortunately, little work has been done in this area. Some interesting ideas have appeared, for example, Suzaki & Hardt [1987] have used a self-tuning controller to regulate the width of a weld bead in a robot welding project. The nice feature of this project is that self-tuning control contributed significantly to the exercise in that the process is noisy and has dynamical characteristics which are difficult to model analytically. In this context the adaption process has positive benefits in that it is difficult to see how the bead width could be effectively controlled without some adaptive mechanism.

The Suzaki & Hardt application is a specific example of a general approach whereby robot sensor data is incorporated in an adaptive framework. Indeed, the challenge for the future will lie in the development of multi-sensor self-adaptive algorithms which can provide the means of adaption to environmental and work-piece fluctuations in sensor based robot control.

5. POSTSCRIPT - SELF-ADAPTIVE OPPORTUNITIES

This postscript was written after the workshop. In this context it is possibly the most (or indeed only) relevant section of this article. In the course of the workshop a number of contributors and participants demonstrated tasks and problems which occur in the handling of limp materials. The issues which seemed of greatest importance in these applications were associated with:-

(a) Sensing in the difficult and ill-structured environment associated with limp material handling, and

(b) The uncertainty associated with the shape, layout and physical characteristics of the workpieces used in limp material handling (e.g: leather, cloth, thread).

The problem of sensing is a major area which cannot be addressed using adaptive techniques. Indeed, unless appropriate sensors exist then self-adaptive methods cannot function, since by their nature self-adaptive methods depend upon sensor data in order to compensate for uncertainty.

In fact the general topic of compensation for uncertainty emerged as the most promising area in which self-adaption may aid in limp material handling. In order to motivate this assertion and to indicate other possible areas of benefit we will reformulate the self-tuning control diagram of figure 1a with limp material handling specifically in mind. The reformulated control scheme will then be used to hypothesize self-adaptive solutions to two application problems which emerged in the workshop presentations. The solutions proposed will be entirely hypothetical in the sense that the approaches which are suggested have not been researched and may not even be possible. The intention however is to demonstrate a methodology whereby self-adaption may be applied.

Figure 5 shows the main elements of a self-tuning system rearranged to suit a discussion on adaption in sensory control of limp material handling. In this context, it is worth recalling that the key elements of an adaptive system are.

(i) the recursive estimator
(ii) the control synthesis (or design block) and
(iii) the control action calculator.

In almost all cases the control action calculation is based upon straightforward techniques and will not therefore be discussed here. The opportunities for innovation exist in the recursive estimator and the control law synthesis. Consider first the recursive estimator; the function of this is to build a model of the process behaviour based upon sensor data. In this vein, another way of describing the recursive estimator is as a model-based sensor signal processor. The function of this signal processing is to fuse the information from the sensors and yield the parameters of a model which represents the sensory data. As indicated in the Figure 5 the resultant model can be used for control synthesis or for process diagnostic studies. This diagnostic use will be referred to again in sections 5.1, 5.2, below.

5.1. Hypothetical Applications - 1. Shoe Leather Machining and Glue Application

A possible application of self adaptive methods in shoe manufacture concerns the fixing of the shoe upper to the sole. This was described in the presentation by David Reedman in which two tasks were described in preparing for the mating of the sole and the upper. These were,

a) automatic machining of the sole bottom in order to prepare a surface suitable for gluing, and

b) the automatic application of a continuous film of adhesive to the sole bottom.

This second task involves tracking of the sole contour (possibly combined with control of the adhesive application) based upon a measurement of the width of the film of glue deposited on the sole bottom. One can draw an analogy between the welding example mentioned in section 4 and the application of adhesive. In particular, provided sensors are available for contour sensing and adhesive monitoring, the self adaptive ideas should transfer straightforwardly from the welding to the shoe glueing application.

The task a) mentioned above could be ammenable to a self-tuning solution if the problem can be described as machining off a certain thickness of leather. Furthermore, if the main problem concerns the uncertainty in the resistance of the leather to machining then a self-tuning system could be proposed which effectively estimates the machining rate. To be specific, a model is proposed which relates the force F(t) applied to a machining tool and the rate of movement of the tool in the direction of the workpiece, v(t). For example, the simplest model form would be

$$v(t) = b_o\, F(t) \tag{10}$$

Thus with sensors available to measure v(t) and F(t) it would be possible to recursively estimate the parameter b_o. The resultant estimate could then be used (as suggested in Figure 5) as a diagnostic indicator of (say) tool wear and

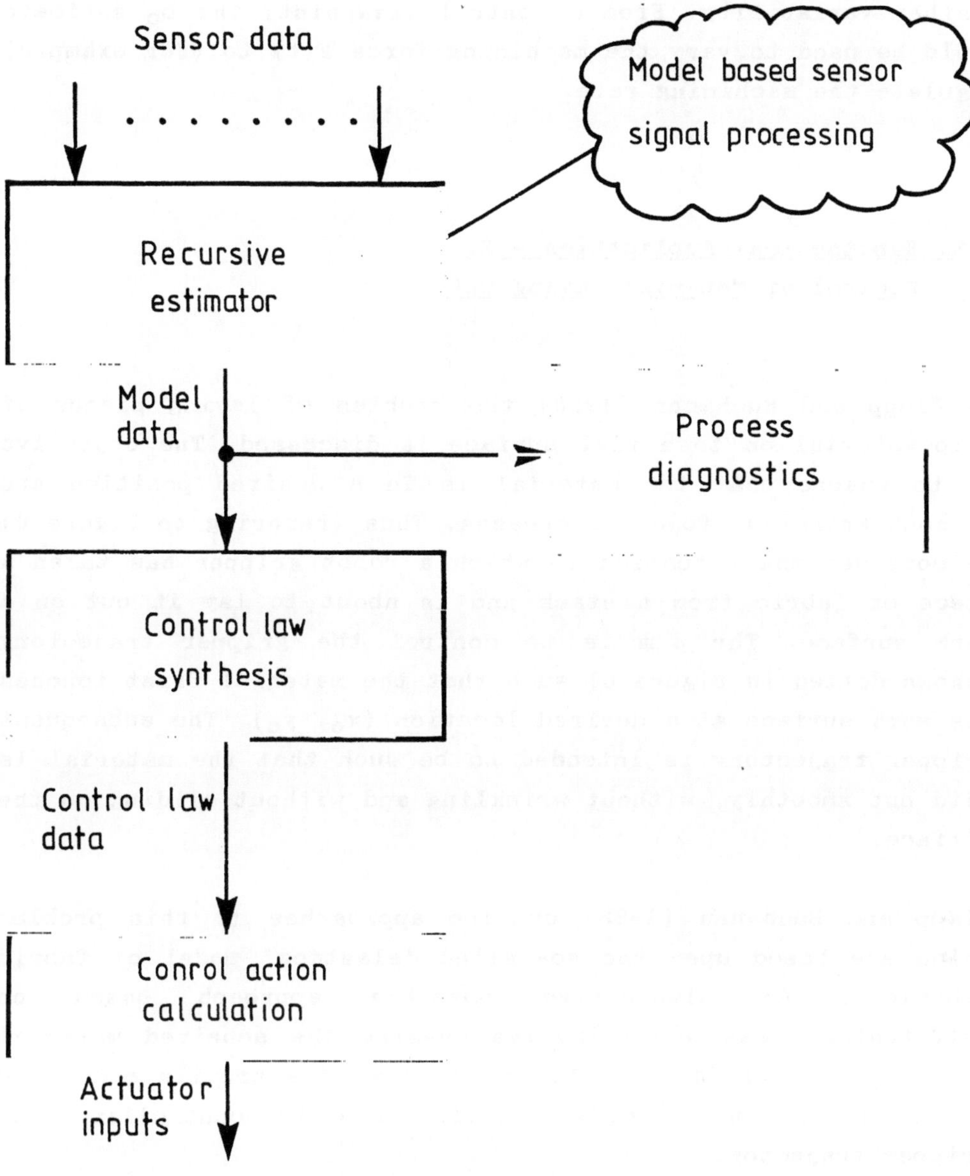

Figure 5. The self-tuning cycle, reformulated for the sensory control of limp materials.

leather variability. From a control viewpoint, the b_0 estimate could be used to vary the machining force F(t) to (for example) regulate the machining rate.

5.2. Hypothetical Application - 2. Control of Material Laying Out

In Clapp and Buchanan [1988] the problem of laying pieces of limp material on to a flat surface is discussed. The objective is to ensure that the material is in a desired position and without wrinkles, folds or creases. Thus (refering to Figure 6) we consider the situation in which a robot gripper has taken a piece of fabric from a stack and is about to lay it out on a work surface. The aim is to control the gripper trajectory (shown dotted in Figure 6) such that the material first touches the work surface at a desired location (x_0, y_0). The subsequent gripper trajectory is intended to be such that the material is laid out smoothly, without wrinkling and without sliding on the surface.

Clapp and Buchanan [1988] outline approaches to this problem which are based upon the so-called 'elastica' model of fabric behaviour. An alternative possible approach based on self-tuning ideas would involve sensing the acheived material side profile, estimating the parameters of a profile model and using the estimated model coefficients for controlling the gripper trajector.

More specifically, we might hypothesize a quadratic form for the material profile as follows:-

$$y(t) = y_0 + \alpha(x(t) - x_0)^2 \qquad (11)$$

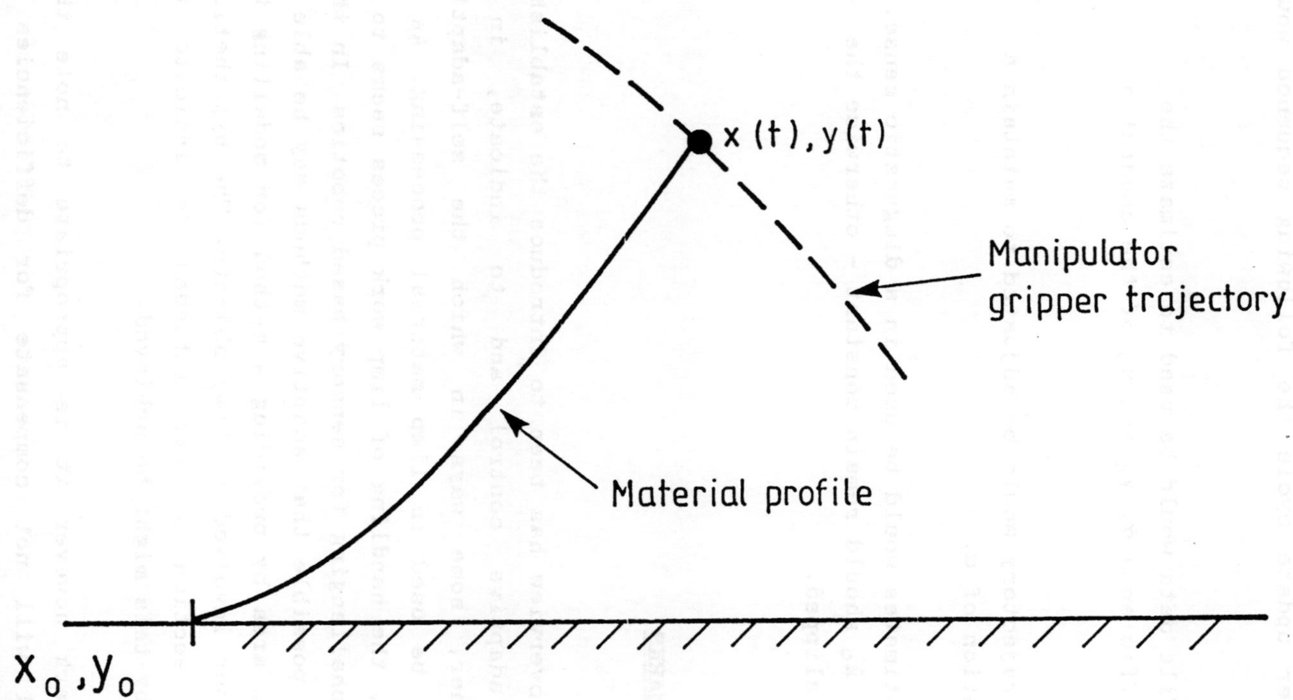

Figure 6. Laying out of limp material.

At each controller update cycle the following sequence would occur.

a) The side profile data would be used to estimate the estimated coefficients $\hat{\alpha}$, $\hat{y}_0$ and $\hat{x}_0$ of the quadratic model.

b) The gripper trajectory would be adjusted to maintain a desired evolution of α.

c) The $\hat{y}_0$, $\hat{x}_0$ estimates would be used in a diagnostic sense. Specifically, $\hat{x}_0$ should remain constant - otherwise the material has slipped.

6. CONCLUDING REMARKS

The aim of this overview has been to introduce the established procedures of adaptive control and to indicate, in a speculative manner, some ways in which the self-adaptive philosophy might be used in limp material processing. As an application area, the handling of limp work pieces seems to be one of the most challenging for sensory based robotics. In this connection it is possible that adaptive methods may be able to contribute to the area by providing a method for modelling the uncertain behaviour involved in limp objects. The hypothetical applications of section 5 were intended to indicate the philosophy whereby this might be achieved.

As a final remark however it is appropriate to note that adaptive control will not compensate for defficiencies in engineering. Indeed, despite its powerful capabilities adaptive control can only serve as an adjunct to good basic engineering.

7. ACKNOWLEDGEMENT

The author acknowledges the contributions and discussions with Din Wahab and Dino Lelic concerning adaptive robot control. The material in section 5 was compiled in consultation with David Reedman of British United Shoe Machinery Ltd and Professor David Buchanan of North Carolina State University. I am grateful for their collaboration but remain responsible for any errors or technical misjudgements which may exist in section 5.

8. REFERENCES AND BIBLIOGRAPHY

ALLIDINA, A.Y. & HUGHES, F.M.: *"Generalized Self-Tuning Controller with Pole-Placement"*, Proc. IEE, 127, Pt.D, 1, pp 13-18, 1980.

ANDERSON, B.D.O., BITMEAD, R.R., JOHNSON, C.R. & KOKOTBIV, P.V.: *"Stability of Adaptive Systems: Passivity and Averaging Analysis"*, MIT Press, 1986.

ANWAR, G., TOMIZUKA, M., HOROWITZ, R., & KUBO, T.: *"Experimental Study on Discrete Time Adaptive Control of an Industrial Robot Arm"*, 2nd IFAC Workshop on Adaptive Systems in Control and Signal Processing, Lund, Sweden, 1986.

ASTROM, K.J.: *"Adaptive Feedback Control"*, Proc. IEEE, 75, 2, pp 185-217, 1987.

ASTROM, K. & WITTENMARK, B.: *"On Self-Tuning Regulators"*, Automatica, 9, pp 185-199, 1973.

BASTIN G., GUILLAUME, A.M. & CAMPION, G.: *"Discrete Time Self-Tuning Computed Torque Controller for Robotic Manipulators"*, Proc 26th IEEE C.D.C., Los Angeles, 1987.

BIERMAN, G.J.: *"Factorization Methods for Discrete Sequential Estimation"*, Academic Press, 1977.

CLAPP, T.G. & BUCHANAN, D.R.: *"Limp Materials Research at North Carolina State University"*, Proceedings of this Workshop, Pisa, Italy, 1988.

CLARKE, D.W., MOHTADI, C. & TUFFS, P.S.: *"Generalized Predicitve Control"*, Automatica, 23, 2, pp 137-148, 1987.

CLARKE, D.W. & GANTHROP, P.G.: *"Self-Tuning Controller"*, Proc. IEE, 122, 9, pp 929-934, 1975.

DUBOWSKY, S. & DESFORGES, D.T.: *"The Application of Model Reference Adaptive Control to Robotic Manipulators"*, J. of Dyn. Syst., Meas. and Control (ASME), Vol. 101, pp 193-200, 1979.

ELLIOT, H., DEPKOVICH, T., KELLY, J. & DRAPER, B.: *"Non-linear Adaptive Control of Mechanical Linkage Systems With Application to Robotics"*, Proc. of Joint Auto. Contr. Confr., pp 1050-1055, 1984.

FREUND, E.: *"Fast Non-Linear Control with Arbitrary Pole-Placement for Industrial Robots and Manipulators"*, The International Journal of Robotic Research, 1, 1, pp. 65-78, 1982.

GENTLEMAN, W.: Journal of Inst Maths Applications, 12, pp. 329-336, 1974.

GOODWIN, G.C. & SIN, K.S.: *"Adaptive Filtering, Prediciton and Control"*, Prentice Hall, 1984.

HOROWITZ, R. & TOMIZUKA, M.: *"Discrete Time Model Reference Control of Mechanical Manipulators"*, Computer in Engineering, Vol.2., Robots and Robotics ASME, pp 107-112, 1982.

KARNY, M. HALONSKOVA, A., BOHM, J., KULHAVY, R. & NEDOMA, P.: *"Design of Linear Quadratic Adaptive Control"*, Kibernetika, 21, pp 3-97, 1985.

KOIVO, A.J. & GUO, T.H.: *"Control of Robotic Manipulator With Adaptive Controller"*, Proc. of the 20th IEEE Confr. on Decision and Control pp 271-276, San Diego, 1981. (see also Trans, IEEE, AC-28, 2, pp.162-171, 1983).

KRAUS, T.W. & MYRON, T.J.: *"Self-Tuning PID Controller Uses Pattern Recognition Approach"*, Control Engineering, June, 1984.

KUČERA, V.: *"Discrete Linear Control"*, Academia Praha, 1979.

LANDAU, I.D.: *"Adaptive Control - The Model Reference Approach"*, Marcel Dekker, 1979.

LAWSON, C.L. & HANSON, R.J.: *"Solving Least Squares Problems"*, Prentice Hall, 1974.

LEININGER, G. & WANG, S.: *"Pole Placement Self-tuning Control of Manipulators"*, IFAC Symp. on CAD of Multivariable Technologycal Systems, Sept. 15-17, West Lafayette, Indiana, USA, 1982.

LELIC, M.A. & WELLSTEAD, P.E.: *"Generalized Pole-Placement Self-Tuning Controller, Part 2: Application to robot manipulator control"*, Int. J. Control, 46, 2, pp 569-601, 1987.

LELIC, A.M. & ZARROP, M.B.: *"Generalized Pole-Placement Self-Tuning Controller"*, Int. J. Control, 46, 2, pp 547-568, 1987.

LJUNG, L. & SODERSTROM, T.: *"Theory and Practice of Recursive Estimation"*, McGraw Hill, 1983.

MURPHY, S.M. & SARIDIS, G.N.: *"Experimental Evaluation of Two Forms of Manipulator Adaptive Control"*, Proc, 26th, IEEE, C.D.C., Los Angeles, 1987.

NORTON, J.: *"Introduction to System Identification"*, Academic Press, 1986.

PRAGER, D.L. & WELLSTEAD, P.E.: *"Multivariable Pole-Assignment Self-Tuning Regulators"*, Proc. IEE. 128, Pt. D., 1981.

SEBORG, D.E., EDGAR, J.F. & SHAH, S.L.: *"Adaptive Control Strategies for Process Control: A Survey"*, A.I.ChE Journal, 32, 6, 1986.

SIDMAN, M.D. & FRANKLIN, G.F.: *"Adaptive Control of a Flexible Structure"*, Proc 2nd IFAC Wokshop on Adaptive Systems in Control and Signal Processing, Lund, Sweden, 1986.

SLOTINE, J.J.E. & LI, W.: *"Adaptive Strategies in the Cooperative Control of Multiple Manipulators", Proc Acc, Minneapolis, 1987.*

SLOTINE, J.J.E. & WEIPING, L.I., *"Adaptive Manipulator Control-Parameter Convergence and Task-Space Strategies"*, ACC Proceedings, pp 828-835, Minneapolis, USA, 1987.

SLOTINE, J.J.E. & COETSEE, J.A.: *"Adaptive Sliding Controller Synthesis for Non-Linear Systems"*, Int. J. Control, 42, 6, 1986.

SUZUKI, A. & HARDT, D.: *"Application of Adaptive Control to In-Process Weld Geometry Regulation"*, Proc, ACC, Minneapolis, 1987.

WAHAB, W. & WELLSTEAD, P.E.: *"Self-Tuning Control of a Robot Manipulator Arm"*, Int. J. Prod. Research (to appear February 1989).

WAHAB, W. & WELLSTEAD, P.E.: *"Multivariable Self-Tuning Controller for Robot Servo Applications"*, 2nd IFAC Workshop on Adaptive Systems and Signal Processing, Lund, Sweden, 1986.

WAHAB, W., *"Adaptive Control of Robot Manipulator"*, Ph.D. Thesis, Control Systems Centre, UMIST, Manchester, UK, 1985.

WELLSTEAD, P.E., WAGNER, G.R. & CALDAS PINTO, J.R.: *"Two-Dimensional Adaptive Prediction, Smoothing and Filtering"*, Proc. IEE, 134, Pt.F, 3, 1987.

WELLSTEAD, P.E., & ZARROP, M.B.: *"Self-Tuning Regulators: Non-Parametric Algorithms"*, Int. J. Control, 37, 4, pp 787-807, 1983.

WELLSTEAD, P.E., PRAGER, D.L. & ZANKER, P.M.: *"Pole-Assignment Self-Tuning Regulator"*, Proc. IEE, 126, 8, pp 781-787, 1979.

WIDROW, B. & STEARNS, S.D.: *"Adaptive Signal Processing"*, Prentice Hall, New Jersey, 1985.

WITTENMARK, B.: *"Self-Tuning Predictor"*, IEEE Trans. AC-19,

YUH, J.: *"Effect of Unmodelled Flexible Dynamics in Discrete Time MRAS of Flexible Single Link Robot"*, Proc 26th, IEEE, C.D.C., Los Angeles, 1987.

ZARROP, M.B. & FISCHER, M.: *"Reduced-Variance Pole-Assignment Self-Tuning Regulation"*, Int. J. Control, 42, 5, pp 1013-1033, 1985.

Fault Tolerant Microprocessor Systems for Industrial Control

C. Preece

School of Engineering and Applied Science
University of Durham
Durham DH1 3LE U.K.

1. Introduction

Current and proposed developments in the automation of the clothing and footwear industries will involve the replacement of manual operations by automatic machines. This automation will be achieved by novel mechanical systems combined with computer control. Control of the automation process will be achieved by microcomputer based systems linked to electromechanical actuators and sensors through electronic interfacing circuitry.

In many areas of automation, analogue control is being superceded by digitally based controllers. This development brings with it greatly enhanced capabilities for sophisticated control of plant, and the flexibility to change control algorithms at low cost. Hardware architecture can be standardised, while software provides the customising element in the design.

A number of general purpose digital controllers are now available on the market, each providing the user with a range of control functions which can be programmed using a simple symbolic language. For many applications, however, it is still necessary to design dedicated equipment.

If digital microprocessor based systems are used in applications where high levels of plant availablilty are important then it is essential to consider those factors which influence reliability, some of which relate to properties of the system itself, and some to the environment in which the system operates.

2. Reliability considerations

Hardware reliability can be determined by analysing the failure rates of discrete electronic components. Each component has a failure rate which is assumed to be constant over its working life. This leads to an exponential reliability function. Values for failure rates are obtained by reference to sources of reliability data such as the U.S. Department of Defense Handbook [1]. Here, each component failure rate can be determined from a formula based on its physical and operational cahracteristics. Failure rates can be determined from this information for complete electronic systems.

The design of a system can be optimised by the use of Fault Mode and Criticality Analysis (FMECA) and Fault Tree Analysis (FTA) techniques. The overall failure rate can be used to determine the Mean Time to Failure (MTTF) of the system, which in the simple case is its reciprocal.

NATO ASI Series, Vol. F 64
Sensory Robotics for the Handling of Limp Materials
Edited by P. M. Taylor

3. Fault Tolerant Systems

Once the overall failure rate, and MTTF, for a system is determined, further improvements in reliability can only be achieved by the implementation of methods to overcome the effects of a fault. A "Fault tolerant" system is defined as one which can continue its designed function in the presence of a fault. If a permanent fault occurs in a system, the function of the system will fail unless an alternative method of achieving the output exists within the system itself. The traditional approach to this problem has been to introduce some form of redundancy into the design.

Redundancy can take the form of duplicate hardware, leading to stand-by systems where a faulty controller is replaced by a duplicate. Redundancy can also be introduced by replication of the software and the recalculation of data using alternate algorithms. Multiple hardware redundant systems such as TMR (Triple Modular Redundant) are well known, and have been implemented in digital as well as analogue systems. Higher orders of hardware redundancy have been used in aerospace applications in recent years.

Redundancy can mask the effect of a fault and provide fault tolerance in the system. It can provide coverage for both permanent and transient faults at a cost of increased complexity in the design.

However other methods of tackling the problems of transient faults are available to the designer, and these are described below.

4. Transient faults

The environment of a garment or footware factory is generally one in which a large number of machines operate in close proximity. Such an environment is particularly unsympathetic to electronic and microprocessor based systems. Electrical interference in a computer circuit can cause signal corruption within the computer itself, leading to either data corruption or erroneous execution of the program. In this case no hardware damage is caused, but the data errors can lead to complete failure of the computer program, and so to the failure of the whole control system. This class of fault is termed a "Transient Fault", and is generally caused by an event external to the system. If the data errors can be detected and corrected then the possibility of complete system recovery exists.

In industrial controllers the likelihood of a transient fault has been estimated to be many times greater than a permanent fault. This type of fault occurs in equipment subject to mains interference, and electromagnetic radiation. Mechanical vibration can also be a source of transient faults due to intermittent contacts on plugs and connectors.

If we take the mains transient as a typical example of externally generated interference, if this finds its way onto the power rails of the electronic circuits then a short transient will be seen on the supply. In an analogue system this might result in an attenuated glitch on the output signals of the controller. In most cases the system would survive unscathed. The same effect on a microprocessor based system can be much more serious. A transient on the supply can produce erroneous signals on the bus-bars leading to corruption of the contents of the dynamic registers and of memory contents.

A worst-case condition exists when the program location counter contents are corrupted leading to a random jump in the program to an undefined memory address. Such a jump may cause the program to halt, and all output function to be lost.

This is an example of the sort of problem which a transient fault can produce, and in automation applications it is important to arrange for some recovery process to be initiated in this situation.

5. Fault tolerance

There is a range of techniques which can be used to provide fault tolerance in a digital controller. Each technique provides coverage for a particular class of fault. More than one technique may be implemented in a particular controller design, based on the details of the controller architecture, the type of function the controller is asked to perform, and details of the plant being controlled.

In an earlier paper [2] fault tolerant techniques were shown to comprise three distinct types. These are summarised in Figure 1.

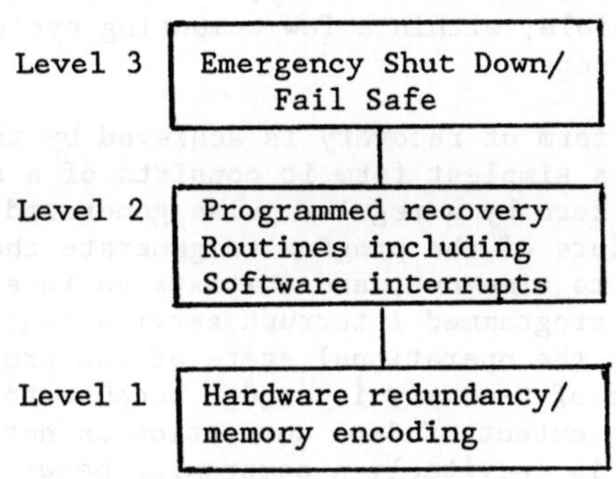

Figure 1 Structured Levels of Fault Tolerance Techniques

Level 1 techniques are those that can be implemented to provide protection from fault conditions in a way that is transparent to the designer. They can be fitted to the hardware as a standard enhancement and are not application dependent. An example of such a technique would be the use of a Hamming Code checking circuit to correct single bit errors in memory.

Level 2 techniques require a knowledge of the software and the application of the system. They must be implemented by the system designer. This is the class of technique illustrated in the examples in section 7 of this paper.

Level 3 techniques are highly application dependent. An example would be a watchdog circuit initiating a safe shut down in the event of a major control failure.

The concept of "levels" is useful to understand the mechanism of the recovery process. Faults should ideally be corrected at Level 1, and only if Level 1 methods fail do Level 2 techniques come into play. When Level 2 methods fail then Level 3 techniques take over to provide a crude but safe control of the plant.

The implementation of Fault Tolerance is generally achieved by a combination of methods at the three Levels defined above.

6. Transient Fault Tolerance

A major problem for microprocessor based controllers is the corruption of data storage and of central processor registers caused by transient disturbances. Power supply transients in particular can cause corruption of the program location counter register leading to a jump in the program. This occurence generally leads to failure of the complete system. Because of the speed of execution of microprocessors, an erroneous jump of this sort can lead to massive corruption of data during subsequent erroneous execution. It is therefore important in considering any mechanism of recovery, to ensure that this is activated as quickly as possible, within a few computing cycles, to minimise the subsequent data errors.

A well known form of recovery is achieved by the so-called "Watch Dog Timer". In its simplest form it consists of a counter circuit which is reset to zero by a regular pulse generated by the running program. Any failure of the program to generate the reset pulse will cause the counter to time-out, and initiate an interrupt to the processor. A pre-programmed interrupt service routine will then attempt to restore the operational state of the program and restart the system. This form of recovery is crude, because the fault type is not identified and the extent of data corruption is not known. The time-out interval is inevitably a compromise between a low value to limit subsequent erroneous execution and a realistic rate of reset pulse generation. More complex watch-dog circuits have been developed where a trace of the program execution is monitored, and any abnormal computation path is flagged. Watch-dog circuits are particulary useful for transient fault protection, as recovery, after a major reset, can be complete because for this class of fault no hardware failure is involved. However, the method relies on the service routine having the capability to restart the process satisfactorily and safely. The routine is therefore specific to the particular process being controlled. In systems where a restart is undesirable or impossible, the routine can be programmed to achieve an ordered shut down or "graceful degradation" of the process to a safe operational state.

In general, watch-dog circuits initiate interrupts or "exceptions" in the microprocessor, and as such are independent of the type of microprocessor being used. Each type of microprocessor however, has internal features which make the handling of interrupts very different between types. In particular some microprocessors have internally generated interrupts which are triggered by specific operational states of the chip. These type-specific features can be used to enhance the probability of recovery following a transient disturbance, and their use provides an additional method of improving fault tolerance. The following section illustrates an example of the difference between microprocessor types; for full details of the methods used the reader is referred to references [3] and [4].

7. Comparison of microprocessor systems

We assume that a microprocessor based digital controller is to be used as part of an automation system, and that this system is to be operated in an industrial environment subject to transient disturbances. If we consider the case of corruption of the program location counter mentioned above, this will lead to a jump to a random location in memory. Subsequent execution at this new location will result in one of the five outcome classes illustrated in Figure 2. Of these outcomes, only one, that of "Restart", gives the possibility of controlled recovery.

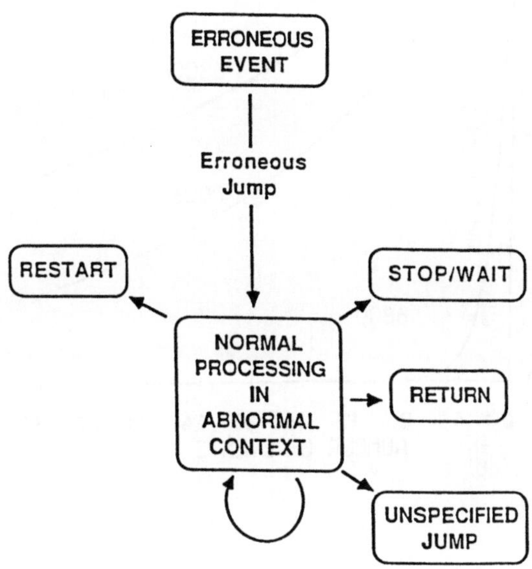

Figure 2. Microprocessor Erroneous Execution Model

The probability of each of the five outcomes can be calculated for each microprocessor type; the probability of the Restart outcome giving a statistical measure of the fault tolerance of the system. The restart probability can be enhanced by careful design of the system, using the particular features of each processor type. Details of the enhancement techniques are given in Reference [3].

Graphs of the Reliability (probability of recovery) for a range of processors each running the same control program and using the same memory area are shown in Figure 3. The examples are taken from

Reference [4]. It can be seen that there is wide difference between the types of microprocessors. The corresponding MTTFs range from 39 days for the M6800 to 13590 years for the M68020 system.

The factors which contribute to the difference in the calculated results are:

i) The addressing range of the processor
ii) The number of software generated interrupts (restarts) in the instruction set
iii) The internally generated interrupts of the processor

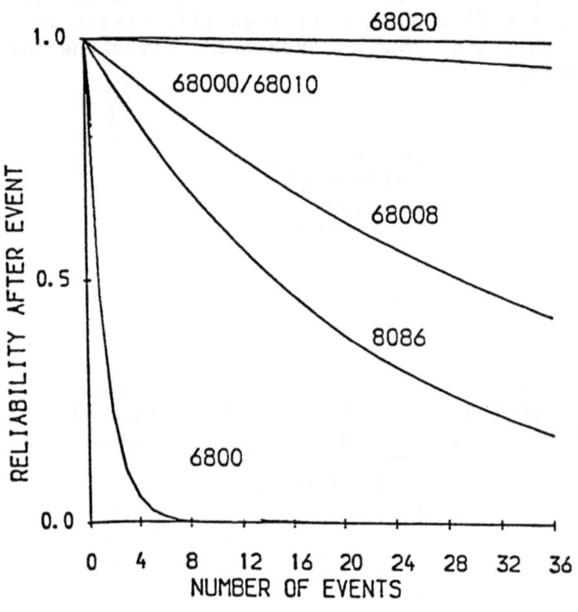

Figure 3. Differences in the Reliability of various microprocessor types

With attention to the integrated design of both hardware and software, the programmer can enhance the recovery capability of the system by utilising the particular features of each microprocessor. The method has the attraction of requiring only a modest increase in the installed memory of the system to contain the extra software involved.

8. Conclusions

The environment in which automation equipment operates in the garment and the footware industries is liable to be electrically noisy. Transient events induced by this environment will cause malfunction in computer based equipment. It is important to recognise the likelihood of this type of malfunction and take measures at the design stage to introduce fault tolerance into the design. A range of techniques are available which can provide coverage for many of the likely fault classes, some of which are of general application. Others rely for their effectiveness on the characteristics of particular microprocessor systems. Techniques have been described which enable the improvement in the recovery probability to be achieved for a variety of microprocessors.

Enhanced fault tolerance in industrial systems gives improved availability of plant and shorter down times. As new automatic systems are introduced into the industry, and as greater reliance is placed on their performance as an integral part of the production flow, the requirement for fault tolerance will increase.

9. Acknowledgements

The author would like to acknowledge the work of his post-graduate colleagues, in particular Dr. R.G. Halse, and G.A.S. Wingate. He would also like to express appreciation to British Gas Engineering Research Station, Newcastle-upon-Tyne, England and the U.K. Science and Engineering Research Council for their support.

10. References

1. U.S. Department of Defense. MIL-HDBK-217-E. 1986.

2. Preece C., Pearson.J.C., and Halse.R.G. The introduction of fault tolerance into digitally controlled gas regulators Proc. Int. Gas Research Conf. London 1983. Paper A11-83. 10pp.

3. Halse R.G. and Preece.C. Erroneous execution and recovery in microprocessor systems. IEE Software and Microsystems Vol 4. No.3. pp.63-70. June 1985.

4. Wingate G.A.S. and Preece.C. Transient fault recovery assessment in 8 and 16 bit microprocessor based controllers in embedded systems. Microprocessing and Microprogramming Vol.24 pp.775-782, 1988

Round Table discussion: Planning and Control: Trends and future Research

Session Chairman: Henry Seesselberg, F.I.T., New York
Secretary: Dr. S Harlock

The session sought to answer the question "Do we have the necessary planning and control systems to solve the problems that may be envisaged in automating clothing manufacture?"

To provide a focus for the discussion, a typical automation problem that confronts the clothing industry was posed as follows:

Given: (i) An automatic pocket-setting machine

(ii) Shirt panel and pocket components supplied by an overhead conveyor.

Task: Devise a solution to automatically load the shirt panel and pocket with the machine and remove completed assembly and replace in overhead hanger.

Assumptions:

(i) Suitable robots and sensors are available

(ii) "Comfortable" funds are available

(iii) 3-year development program.

It was agreed that a distributed hybrid control system would most probably be required. Factors to be taken into account in deciding on the requirements of the system would be:-

(i) Sequence of events

(ii) Potential errors

(iii) Collision avoidance.

Within the context of potential errors consideration was given to inspection and diagnostics and the dependence of decisions taken on market forces, viz: are error recovery costs greater than garment costs? was stressed. The importance of assessing trends in diagnostic analysis was pointed out to identify random and systematic effects: it was considered doubtful whether adaptive control systems could be developed that would obviate the need for diagnostics. The appropriate solution would be to assess the number of error recovery cycles and use this to give diagnostic information.

Under the minimal constraints imposed it was generally agreed that this problem, and similar problems, were soluble with existing control technology.

Additional constraints of processing time and cost were imposed. (6 second cycle time, system cost $50K). The strategy to solve the problem was to use general purpose robots and systems, and refine to give a solution or define requirements for research to ensure a solution.

However, the whole concept of trying to effect solutions on current machinery was brought into question. A more sensible approach to this and

NATO ASI Series, Vol. F 64
Sensory Robotics for the Handling of Limp Materials
Edited by P. M. Taylor

other problems would be to redesign the method and plan for the whole function rather than replace or enhance an existing function.

CONCLUSION

Using the problem as a vehicle for assessing strategy and the level of technology available, it was agreed that the planning and control technology is currently available but not at an affordable price. More research is required to reduce the cost of the technology to make solutions economically viable.

CHAPTER V

ROUND-UP

Final Round Table Discussion

Chairman: Dr. C. Preece
Secretary: Mr. G. Monkman

Six points were displayed on the blackboard:

1. Communication
2. Definitions
3. Anthropormorphic
4. Looking ahead
5. Are we doing the right things?
6. Are we all here?

Dr. Preece started with the last question first "Is *everyone here who should be here? - if not why not?*"

A brief head count revealed:
1 Garment manufacturer.
4 Machinery manufacturers.
5 Non-University researchers.
20 University researchers.

Dr. Paul Taylor of Hull University stated that far more had been invited but some had declined - particularly small companies. Apparently, this workshop clashed with another (US) textile meeting being held the following week. Details of conference time lengths and other possible reasons why some industrialists had not attended were then mentioned. Some wondered why no composite manufacturers had turned up as they were usually more wealthy in terms of resources etc. We were informed that the Society of Industrial Robotics (Italy) were apparently unaware of the workshop.

The rest of the above points were then investigated, the question "*Are we technology led or application led*" being pursued. Mr. Hubert Blessing (Levi-Strauss) stated the need for basic research to find out what is really required. This was followed by the suggestion that, 'whatever it is, the garment industry will be unable to afford it'. This echoed the view of a previous round table session in that most sensors used in the textile industry were originally intended for other uses. Mr. David Reedman gave the opinion that researchers were becoming too commercially aware to the possible detriment of long term research goals. Mr. Gareth Monkman of Hull University agreed, but doubted whether present political and industrial sentiment, in the UK at least, was capable of such foresight. Most present concluded that long term research had much less chance of being funded than short term research.

Talking about logistics rather than technicalities - are companies being educated? asked Herr. Nollek (IPA Stuttgart). The trend of engineers becoming managers was then discussed. The US citizens present noted that most managers there are MBA's and that accountants are never given training in research justification. It was agreed that a different character of management was required.

It was felt that communications must be improved between researchers in sensing, control and applied robotics, equipment manufacturers and end users. Ways of achieving this were discussed with a future workshop being favoured.

NATO ASI Series, Vol. F 64
Sensory Robotics for the Handling of Limp Materials
Edited by P. M. Taylor

Final Round Table Discussion

Chairman: Dr. G. Preece
Secretary: Mr. G. Monkman

Six points were displayed on the blackboard:

1. Communication
2. Definitions
3. Anthropomorphic
4. Looking ahead
5. Are we doing the right things?
6. Are we all here?

Dr. Preece started with the last question first "*is everyone here who should be here? - if not why not?*"

A brief head count revealed:
1 Garment manufacturer.
4 Machinery manufacturers.
5 Non-University researchers.
20 University researchers.

Dr. Paul Taylor of Hull University noted that far more had been invited but some had declined, particularly small companies. Apparently, this workshop clashed with another [illegible] meeting being held the following week. [illegible] of conferences, time [illegible] and other [illegible] reasons why some manufacturers had not attended were then mentioned. Some wondered why no composite manufacturers had turned up as they were usually more wealthy in terms of resources etc. We were informed that the [illegible] of Industrial Robotics (Italy) were apparently unaware of the workshop.

The rest of the above points were then investigated, the question "Are we technology led or application led" being pursued. Mr. Hubert Blessing (Levi-Strauss) stated the need for basic research to find out what is really required. This was followed by the suggestion that, "whatever it is, the garment industry will be unable to afford it". This echoed the view of a previous round table session in that most sensors used in the textile industry [illegible] were looking for commercially [illegible] to the possible detriment of long term research goals. Mr. Gerry Stephenson of Hull University agreed, but [illegible] and industrial sentiment, in the UK at least, [illegible] funded than short term research.

Talking about logistics rather than technologies in the companies, being necessary [illegible] (IPA Stuttgart). The trend of engineers becoming managers was then discussed. The US citizens present noted that most managers there are MBA's and that accountants are never given training in research justification. It was agreed that a different character of management was required.

It was felt that communications must be improved between researchers in sensing, control and applied robotics, equipment manufacturers and end users. Ways of achieving this were discussed with a future workshop being favoured.

NATO ASI Series, Vol. F 64
Sensory Robotics for the Handling of Limp Materials
Edited by P. M. Taylor

LIST OF SPEAKERS

D.R. Buchanan
College of Textiles
North Carolina State University
Box 8301
Raleigh, NC
USA

A.J. Campilho
Faculdade de Engenharia
University of Porto
Rua dos Bragas
4099 Porto Cedex
Portugal

J. Domey
Electrical Engineering Division
National Research Centre
Ottawa
Ontario KIA OR8
Canada

P. Esquirol
Ecole Nationale Superiere
des Techniques
6 Ave des Clavieres
30107 Ales Cedex
France

B. Gasmi
Dept. d'Etudes et des Recherches
en Automatique
CERT
2 Avenue Edouard Belin, BP 4025
31055 Toulouse Cedex
France

S.C. Harlock
Department of Textile Industries
University of Leeds
Leeds LS2 9JT
UK

M. Kamel
Dept. Systems Design Engineering
University of Waterloo
Waterloo
Ontario N2L 3G1
Canada

R.B. Kelley
School of Engineering
Rensselaer Polttechnic Institute
Troy
New York 12180 - 3590
USA

H. Nollek
IPA (Stuttgart)
Postfach 800 469
Nobelstrasse 12
D-7000 Stuttgart 80
West Germany

F.W. Paul
Department of Mechanical Engineering
Clemson University
Clemson
South Carolina 29634 - 0921
USA

A.J. Padilha
Faculdade de Engenharia
University of Porto
Rua dos Bragas
4099 Porto Cedex
Portugal

P.J. Perkins
SATRA Footwear Technology Centre
Rockingham Road
Kettering
Northants NN16 9JH
UK

C. Preece
School of Engineering and Applied Science
University of Durham
South Road
Durham
UK

M. Sarhadi
School of Engineering and Applied Science
University of Durham
South Road
Durham
UK

H.A. Seesselberg
Fashion Institute of Technology
New York
USA

G.E. Taylor
Department of Electronic Engineering
University of Hull
Hull HU6 7RX
UK

P.M. Taylor
Department of Electronic Engineering
University of Hull
Hull HU6 7RX
UK

F. Vercraene
Ecole Nationale Superiere
des Techniques
6 Ave des Clavieres
30107 Ales Cedex
France

P.E. Wellstead
Control Systems Centre
University of Manchester
Institute of Science and Technology
Sackville Street
Manchester M60 1QD
UK

LIST OF PARTICIPANTS

F. Alcantara
Investronica
Tomas Breton 62
Madrid
Spain

F. Bisson
Necchi Spa
Via Rismordo 78
27100 Pavia
Italy

H. Blessing
Levi-Strauss Inc.
Research and Development Center
900 Dorothy Drive
Richardson
Texas 75081
USA

D. Caldwell
Department of Electrical
and Electronic Engineering
University of Salford
Salford
UK

G. Eisenbeis
Pfaff GmbH
Konigstrasse 154
6750 Kaiserslautern
PO Box 3020
West Germany

G. Monkman
Department of Electronic Engineering
University of Hull
Hull HU6 7RX
UK

U. Nunes
Dept. Eng. Electrotechnica
University of Coimbra
3000 Coimbra
Portugal

D.C. Reedman
British United Shoe Machinery
PO Box 88
Ross Walk
Leicester LE4 5BX
UK

G. Schulz
Institut fur Textil und Verfahrenstechnik
Korschtalstrasse 26
Denkendorf
West Germany

J. Simmons
School of Engineering and Applied Science
University of Durham
South Road
Durham
UK

E. Trento
Prima Electronics S.P.A.
Strada Carignano 48/2
10024 Moncalieri (TO)
Italy

NATO ASI Series F

Including Special Programme on Sensory Systems for Robotic Control (ROB)

Vol. 1: Issues in Acoustic Signal – Image Processing and Recognition. Edited by C. H. Chen. VIII, 333 pages. 1983.

Vol. 2: Image Sequence Processing and Dynamic Scene Analysis. Edited by T. S. Huang. IX, 749 pages. 1983.

Vol. 3: Electronic Systems Effectiveness and Life Cycle Costing. Edited by J. K. Skwirzynski. XVII, 732 pages. 1983.

Vol. 4: Pictorial Data Analysis. Edited by R. M. Haralick. VIII, 468 pages. 1983.

Vol. 5: International Calibration Study of Traffic Conflict Techniques. Edited by E. Asmussen. VII, 229 pages. 1984.

Vol. 6: Information Technology and the Computer Network. Edited by K. G. Beauchamp. VIII, 271 pages. 1984.

Vol. 7: High-Speed Computation. Edited by J. S. Kowalik. IX, 441 pages. 1984.

Vol. 8: Program Transformation and Programming Environments. Report on a Workshop directed by F. L. Bauer and H. Remus. Edited by P. Pepper. XIV, 378 pages. 1984.

Vol. 9: Computer Aided Analysis and Optimization of Mechanical System Dynamics. Edited by E. J. Haug. XXII, 700 pages. 1984.

Vol. 10: Simulation and Model-Based Methodologies: An Integrative View. Edited by T. I. Ören, B. P. Zeigler, M. S. Elzas. XIII, 651 pages. 1984.

Vol. 11: Robotics and Artificial Intelligence. Edited by M. Brady, L. A. Gerhardt, H. F. Davidson. XVII, 693 pages. 1984.

Vol. 12: Combinatorial Algorithms on Words. Edited by A. Apostolico, Z. Galil. VIII, 361 pages. 1985.

Vol. 13: Logics and Models of Concurrent Systems. Edited by K. R. Apt. VIII, 498 pages. 1985.

Vol. 14: Control Flow and Data Flow: Concepts of Distributed Programming. Edited by M. Broy. VIII, 525 pages. 1985.

Vol. 15: Computational Mathematical Programming. Edited by K. Schittkowski. VIII, 451 pages. 1985.

Vol. 16: New Systems and Architectures for Automatic Speech Recognition and Synthesis. Edited by R. De Mori, C.Y. Suen. XIII, 630 pages. 1985.

Vol. 17: Fundamental Algorithms for Computer Graphics. Edited by R.A. Earnshaw. XVI, 1042 pages. 1985.

Vol. 18: Computer Architectures for Spatially Distributed Data. Edited by H. Freeman and G.G. Pieroni. VIII, 391 pages. 1985.

Vol. 19: Pictorial Information Systems in Medicine. Edited by K. H. Höhne. XII, 525 pages. 1986.

Vol. 20: Disordered Systems and Biological Organization. Edited by E. Bienenstock, F. Fogelman Soulié, G. Weisbuch. XXI, 405 pages.1986.

Vol. 21: Intelligent Decision Support in Process Environments. Edited by E. Hollnagel, G. Mancini, D.D. Woods. XV, 524 pages. 1986.

NATO ASI Series F

Vol. 22: Software System Design Methods. The Challenge of Advanced Computing Technology. Edited by J. K. Skwirzynski. XIII, 747 pages. 1986.

Vol. 23: Designing Computer-Based Learning Materials. Edited by H. Weinstock and A. Bork. IX, 285 pages. 1986.

Vol. 24: Database Machines. Modern Trends and Applications. Edited by A. K. Sood and A. H. Qureshi. VIII, 570 pages. 1986.

Vol. 25: Pyramidal Systems for Computer Vision. Edited by V. Cantoni and S. Levialdi. VIII, 392 pages. 1986. *(ROB)*

Vol. 26: Modelling and Analysis in Arms Control. Edited by R. Avenhaus, R. K. Huber and J. D. Kettelle. VIII, 488 pages. 1986.

Vol. 27: Computer Aided Optimal Design: Structural and Mechanical Systems. Edited by C. A. Mota Soares. XIII, 1029 pages. 1987.

Vol. 28: Distributed Operating Systems. Theory und Practice. Edited by Y. Paker, J.-P. Banatre and M. Bozyiğit. X, 379 pages. 1987.

Vol. 29: Languages for Sensor-Based Control in Robotics. Edited by U. Rembold and K. Hörmann. IX, 625 pages. 1987. *(ROB)*

Vol. 30: Pattern Recognition Theory and Applications. Edited by P. A. Devijver and J. Kittler. XI, 543 pages. 1987.

Vol. 31: Decision Support Systems: Theory and Application. Edited by C. W. Holsapple and A. B. Whinston. X, 500 pages. 1987.

Vol. 32: Information Systems: Failure Analysis. Edited by J. A. Wise and A. Debons. XV, 338 pages. 1987.

Vol. 33: Machine Intelligence and Knowledge Engineering for Robotic Applications. Edited by A. K. C. Wong and A. Pugh. XIV, 486 pages. 1987. *(ROB)*

Vol. 34: Modelling, Robustness and Sensitivity Reduction in Control Systems. Edited by R. F. Curtain. IX, 492 pages. 1987.

Vol. 35: Expert Judgment and Expert Systems. Edited by J. L. Mumpower, L. D. Phillips, O. Renn and V. R. R. Uppuluri. VIII, 361 pages. 1987.

Vol. 36: Logic of Programming and Calculi of Discrete Design. Edited by M. Broy. VII, 415 pages. 1987.

Vol. 37: Dynamics of Infinite Dimensional Systems. Edited by S.-N. Chow and J. K. Hale. IX, 514 pages. 1987.

Vol. 38: Flow Control of Congested Networks. Edited by A. R. Odoni, L. Bianco and G. Szegö. XII, 355 pages. 1987.

Vol. 39: Mathematics and Computer Science in Medical Imaging. Edited by M. A. Viergever and A. Todd-Pokropek. VIII, 546 pages. 1988.

Vol. 40: Theoretical Foundations of Computer Graphics and CAD. Edited by R. A. Earnshaw. XX, 1246 pages. 1988.

Vol. 41: Neural Computers. Edited by R. Eckmiller and Ch. v. d. Malsburg. XIII, 566 pages. 1988.

NATO ASI Series F

Vol. 42: Real-Time Object Measurement and Classification. Edited by A. K. Jain. VIII, 407 pages. 1988. *(ROB)*

Vol. 43: Sensors and Sensory Systems for Advanced Robots. Edited by P. Dario. XI, 597 pages. 1988. *(ROB)*

Vol. 44: Signal Processing and Pattern Recognition in Nondestructive Evaluation of Materials. Edited by C. H. Chen. VIII, 344 pages. 1988. *(ROB)*

Vol. 45: Syntactic and Structural Pattern Recognition. Edited by G. Ferraté, T. Pavlidis, A. Sanfeliu and H. Bunke. XVI, 467 pages. 1988. *(ROB)*

Vol. 46: Recent Advances in Speech Understanding and Dialog Systems. Edited by H. Niemann, M. Lang and G. Sagerer. X, 521 pages. 1988.

Vol. 47: Advanced Computing Concepts and Techniques in Control Engineering. Edited by M. J. Denham and A. J. Laub. XI, 518 pages. 1988.

Vol. 48: Mathematical Models for Decision Support. Edited by G. Mitra. IX, 762 pages. 1988.

Vol. 49: Computer Integrated Manufacturing. Edited by I. B. Turksen. VIII, 568 pages. 1988.

Vol. 50: CAD Based Programming for Sensory Robots. Edited by B. Ravani. IX, 565 pages. 1988. *(ROB)*

Vol. 51: Algorithms and Model Formulations in Mathematical Programming. Edited by S.W. Wallace. IX, 190 pages. 1989.

Vol. 52: Sensor Devices and Systems for Robotics. Edited by A. Casals. IX, 362 pages. 1989. *(ROB)*

Vol. 53: Advanced Information Technologies for Industrial Material Flow Systems. Edited by S. Y. Nof and C. L. Moodie. IX, 710 pages. 1989.

Vol. 54: A Reappraisal of the Efficiency of Financial Markets. Edited by R. M. C. Guimarães, B. G. Kingsman and S. J. Taylor. X, 804 pages. 1989.

Vol. 55: Constructive Methods in Computing Science. Edited by M. Broy. VII, 478 pages. 1989.

Vol. 56: Multiple Criteria Decision Making and Risk Analysis Using Microcomputers. Edited by B. Karpak and S. Zionts. VII, 399 pages. 1989.

Vol. 57: Kinematics and Dynamic Issues in Sensor Based Control. Edited by G. E. Taylor. XI, 456 pages. 1990. *(ROB)*

Vol. 58: Highly Redundant Sensing in Robotic Systems. Edited by J. T. Tou and J. G. Balchen. X, 322 pages. 1990. *(ROB)*

Vol. 59: Superconducting Electronics. Edited by H. Weinstock and M. Nisenoff. X, 441 pages. 1989.

Vol. 60: 3D Imaging in Medicine. Algorithms, Systems, Applications. Edited by K. H. Höhne, H. Fuchs and S. M. Pizer. IX, 460 pages. 1990.

Vol. 61: Knowledge, Data and Computer-Assisted Decisions. Edited by M. Schader and W. Gaul. VIII, 421 pages. 1990.

Vol. 62: Supercomputing. Edited by J. S. Kowalik. X, 425 pages. 1990.

Vol. 63: Traditional and Non-Traditional Robotic Sensors. Edited by T. C. Henderson. VIII, 463 pages. 1990. *(ROB)*

Vol. 64: Sensory Robotics for the Handling of Limp Materials. Edited by P. M. Taylor. IX, 343 pages. 1990. *(ROB)*

MIX
Papier aus verantwortungsvollen Quellen
Paper from responsible sources
FSC® C105338

If you have any concerns about our products,
you can contact us on
ProductSafety@springernature.com

In case Publisher is established outside the EU,
the EU authorized representative is:
Springer Nature Customer Service Center GmbH
Europaplatz 3, 69115 Heidelberg, Germany

Printed by Libri Plureos GmbH
in Hamburg, Germany